INTRODUCTION TO WELDING

Ronald F. Gonzales
Automotive Servicing/Welding Instructor
Cypress College

I. G. Edmonds
Author of technical/instructional Books

Canfield Press
San Francisco

A Department of Harper & Row, Publishers, Inc.
New York Evanston London

DEDICATION

To Michael Gonzales

ACKNOWLEDGMENTS

We wish to thank the Welding Department, Cypress College, Cypress, California, for permitting the use of its facilities and equipment for some of the photographs. All photographs, unless otherwise acknowledged, were taken by us.

We would also like to express the deepest appreciation to John Alvin, Lebanon, Oregon; Ryan L. Collins, Clark College, Vancouver, Washington; and John E. Ugar, Portland Community College, for suggestions that eased our task and materially improved the manuscript.

Ronald F. Gonzales
I. G. Edmonds

Editors: Richard Thiel, Eva Marie Strock
Cover Design: Margaret Sanfilippo, W B Associates
Interior Design: Donna Plath, W B Associates
Line illustrations: John Foster

This book was set in 10 point Century Schoolbook by W B Associates.
Printed and bound by Halliday Litograph.

INTRODUCTION TO WELDING

Library of Congress Cataloging in Publication Data

Gonzales, Ronald, 1944—
Introduction to welding

Includes index.
1. Welding. I. Edmonds, I. G., joint author.
II. Title
TS227.G6327 671.5'2 75-26668
ISBN 0-06-453303-4

78 79 10 9 8 7 6 5 4 3 2

PREFACE

In choosing to become a welder, you are making a decision to join one of the world's most useful trades. Welding is also a trade in great demand, and indications are that the demand for good welders will increase significantly in the next few years.

The principles of welding are easy to learn, but welding is an art that requires considerable practice. You may know exactly what to do but cannot quite make your hands, torch, or arc do what you want them to. In the beginning your puddle may run, your filler rod may stick, or your arc may burn a hole right through the base metal you are trying to weld.

These and dozens of other mistakes beginners make are normal. We have all made them, so do not get discouraged. You often learn more through your errors than by doing things right. If you keep practicing, in a surprisingly short time you will be able to lay a weld bead with complete control.

This book is an *introduction* to welding. It does not pretend to teach you all there is to know about the welding craft. The purpose of this text is to help you learn to make simple welds on common metals, enabling you to do a wide variety of professional jobs.

Far too many instructional books seem to have been written for either the author or teachers rather than the student. To avoid that mistake in this book, we asked many students what they would like to see in an instructional book for beginning welders. The answer we got most often was "Make it more simple." To this end, we have kept technical data to a minimum. Some technical data are unavoidable, but all that is not essential to you the beginner has been eliminated to permit easier learning. Later, when you study more advanced techniques, you can learn the additional technical points you will need.

The fundamental information in this book is all you really need to know to become a good basic welder. After you learn these points and techniques, all that remains for you to do is to practice intensively so you can gain proficiency.

Good luck, and we are delighted to have you join us in the world of welding.

CONTENTS

Preface iii

Part I. Welding and Metals 1
- Chapter 1 — This is Welding 2
- Chapter 2 — Personal Safety 18
- Chapter 3 — Something about Metals 30

Part II. Oxyacetylene Welding 49
- Chapter 4 — The Oxyacetylene Process 50
- Chapter 5 — Preparing Metal for Oxyacetylene Welding 70
- Chapter 6 — The Welding Torch 84
- Chapter 7 — Welding Rods 102
- Chapter 8 — Forming the Bead 111
- Chapter 9 — The Effect of Heat on Metal 138
- Chapter 10 — Out-of-Position Welding 150
- Chapter 11 — Brazing and Braze Welding 164
- Chapter 12 — Soldering 181
- Chapter 13 — Oxyacetylene Flame Cutting 194
- Chapter 14 — Special Cutting Techniques 211

Part III. Arc Welding 223
- Chapter 15 — Principles of Arc Welding 224
- Chapter 16 — Arc Welding Electrodes 239
- Chapter 17 — Machine Setup and Striking the Arc 251
- Chapter 18 — Weld Inspection 265

Part IV. Special Applications 277
- Chapter 19 — Art Metal Welding 278
- Chapter 20 — Home Welding 292
- Chapter 21 — Where Do We Go From Here? 307

Index 313

WELDING AND METALS

THIS IS WELDING

A hooded man, braced upon a naked steel girder 40 stories in the air, directs a stream of fire at a connection of steel joints. The fire, burning at about 6300° Fahrenheit (F), is slightly more than half as hot as the surface of the sun itself. As this man works, the intense flame he holds melts the edges of two pieces of steel and a filler rod. The molten metal flows together and fuses the two pieces of steel into a single piece. When the metal cools, the bond is as strong as the original steel.

This hooded man is a welder, an indispensable member of a team that is building a modern skyscraper.

Across town a handyman in his backyard hunches over the broken parts of his lawn mower. He likewise directs a stream of fire at the edges of the broken metal parts. He melts them back together with equipment that cost less than $100, equipment which will save him many times $100 on future repairs. This man, in a sense, is also a welder.

WELDERS ARE ALMOST EVERYWHERE

You will also find welders in diving suits under the sea. They do the seemingly impossible: making fire burn under water. Welders are also under the kitchen sink, repairing broken pipe; building aircraft in huge factories; putting up swings for kids; building refrigerators and cars in assembly plants; or even sculpting metal art in artists' studios. Welders are also busy in garages, repairing wrecks, and in hobby shops, turning out wrought iron planters, bike carriers, and a variety of other interesting projects.

Welders are not everywhere, but you will rarely find a place that is not making some use of the welder's work, including even the home or the office, where the only labor is thinking and shuffling papers. All things, from bedsteads to chairs, filing cabinets to typewriters, have some welding in their makeup.

WELDING AS A CAREER

The widespread use of welding today makes welding a career with an excellent future. Welding can also be adventurous. Although many welders work an 8- to 5-shift and like it that way, you will find welders making sparks fly in far corners of the world, working odd and/or long hours. They weld pipelines in Alaska, build bridges in Asia, repair mining machinery in Africa, and weld broken earth-moving equipment in India. Welders work on space rockets, and the hulls of submarines and ships are the result of the welder's skill.

A welder's job is more than just joining together two pieces of metal. He may be required to reinforce worn parts or restore a broken gear tooth. He can use the welding arc to lay a coating of a harder metal over a softer one to make it wear longer. The welder can weld studs into machinery, saving the time once spent in drilling and tapping holes for studs to be screwed into.

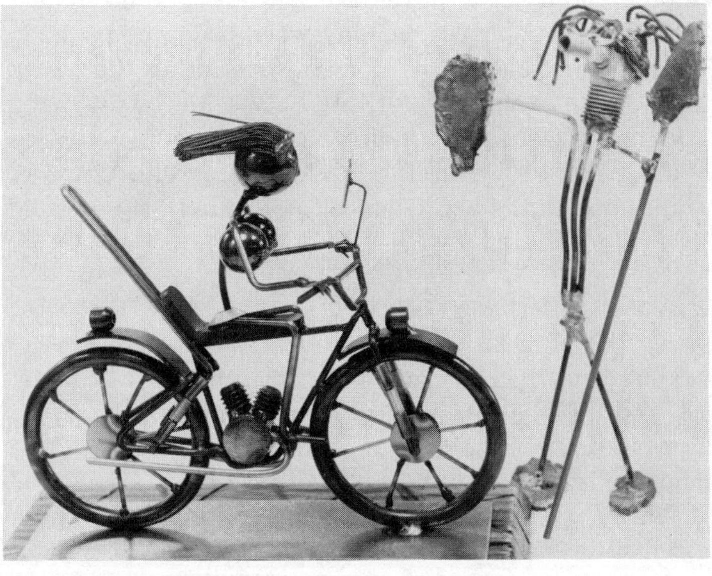

Figure 1-1. Welders may also be artists—or even cartoonists—in metal.

Figure 1-2. The widespread use of welding in industry makes welding a career with an excellent future.

A TOOL FOR NONPROFESSIONAL WELDERS

Many men who never intend to make welding a profession take welding courses to gain a useful skill for their own work. For example, plumbers often can use a welding torch to an advantage. Automobile mechanics frequently have to use welding when doing body work. Farmers can save a lot of money repairing their own tools. These examples illustrate only a few of the many ways a knowledge of welding can serve people engaged in all types of work. If you learn to weld, you will acquire one of the world's most useful skills.

HOW WELDING BEGAN

Welding is as old as man's work with metal. It began at the end of the Stone Age, when prehistoric people discovered that bits of a soft metal like gold could be hammered together to make larger pieces.

Later, probably from watching volcanoes in action, man observed that heat softened the harder metals

Figure 1-3. Japanese swordsmiths still use the traditional forge method of welding to hammer hot iron into blades. This type of welding was used by ancient swordsmiths, armorers, and blacksmiths for centuries.

enough so that he could shape them into tools and weapons. This discovery led to the Bronze Age and then the Iron Age. Metal did not have to be melted to be formed. Hard metals could be hammered into weapons, if they were first heated just enough to make them soft.

Joining or forming metals by heating and hammering is called pressure or *forge welding*. For more than 3,000 years forge welders plied their trade as blacksmiths, armorers, and swordsmiths. They were brawny, strong-armed men who heated iron and steel in forges and then hammered them into swords, horseshoes, plowshares, wheel rims, nails, or any other metal objects they required.

As civilization advanced, there was an increasing need for faster and better ways to join metal. The rivet was utilized. Girders for skyscrapers, boilers for locomotives, plates for ships, and other heavy construction depended upon rivets to join together their parts. The rivets were heated red hot, placed in holes previously bored in the metal, and bradded by hammering. But there was still a need for even more efficient and faster welding methods.

FUSION WELDING

Engineers knew that construction could be done better and faster if metal edges could be melted and fused together. The problem was that iron melts at 2786° F. Such an enormous temperature could be achieved in steel mills, but it could not be produced in flames small enough for a man to use while straddling a girder 50 stories in the air.

The first practical fusion welding was thermit welding, developed in 1895, which had limited application for welding large pieces of metal. Thermit welding is based on the chemical reaction that takes place when aluminum and iron oxide are heated. The aluminum combines with the oxygen from the iron oxide, generating heat that melts the iron. If such a compound is placed in a mold around two pieces of iron—such as railroad rails—and heat is applied, the resulting high temperature melts and fuses the iron.

Figure 1-4. An oxyacetylene setup. The needle valves (G) control the oxygen and acetylene pressure to the torch tip (I). The oxyacetylene flame (J) burns at temperatures ranging from 5500 to 6300° F—hot enough to melt any commercially used metal.

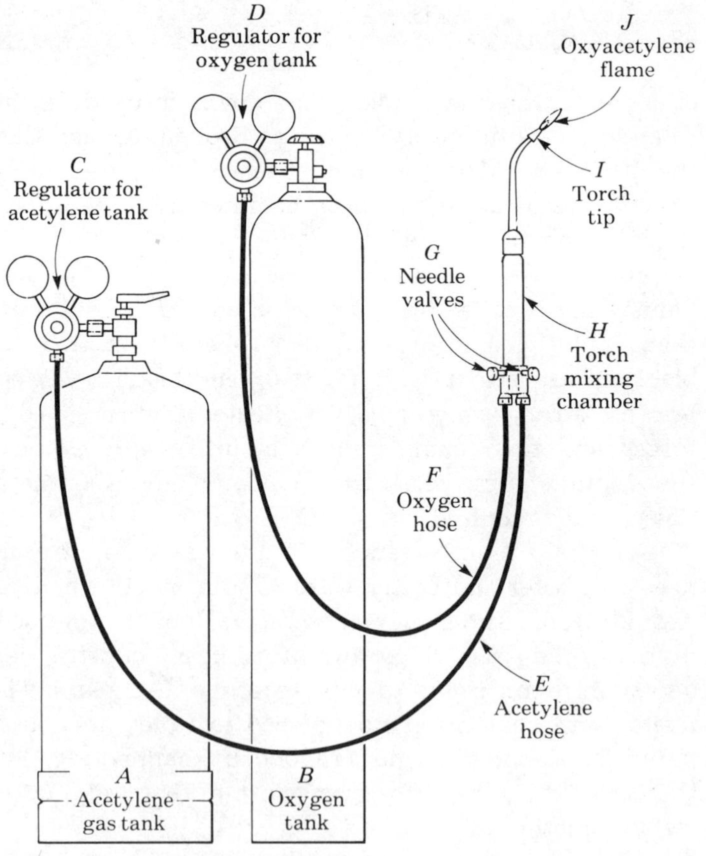

OXYACETYLENE WELDING

The first practical portable welding system for general use was developed in 1901 by two Frenchmen, Fouche and Picard. They burned acetylene gas mixed with pure oxygen in a blow torch.

Acetylene is a poisonous, highly flammable gas created by the chemical reaction of calcium carbide with water. (Calcium carbide is used in pellet form.) Acetylene burns at a temperature of about 6300°F when mixed with oxygen. This is more than sufficient heat to melt any commercial metal in use today.

ARC WELDING

Arc welding is actually older than oxyacetylene welding, but it was more difficult to perfect. Oxyacetylene went into general use much sooner.

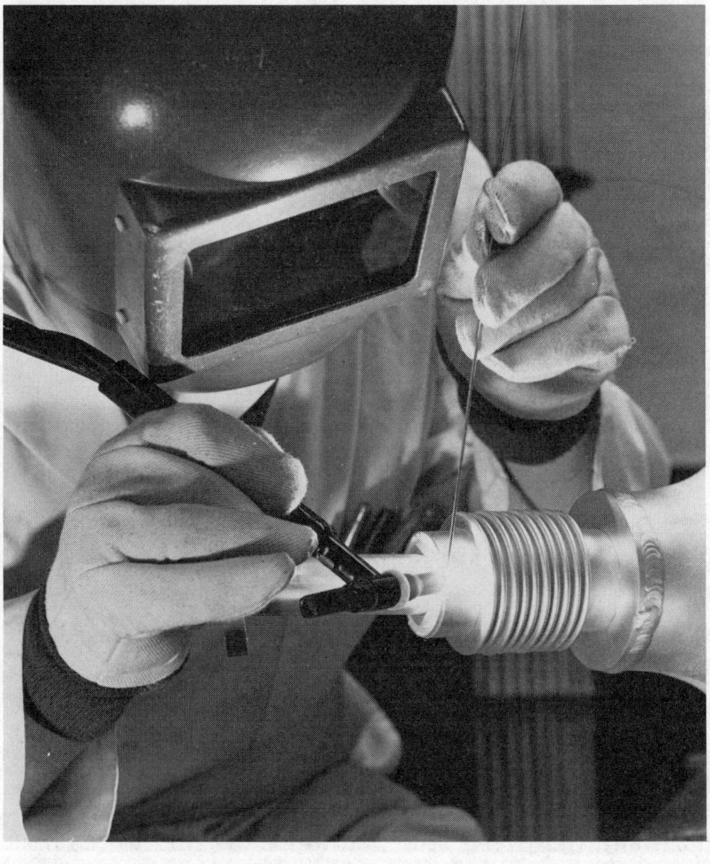

Figure 1-5. Arc welding uses an electric arc to melt the base metal and filler rod. The welder shown here is using a system called Heliarc to weld an aircraft part.

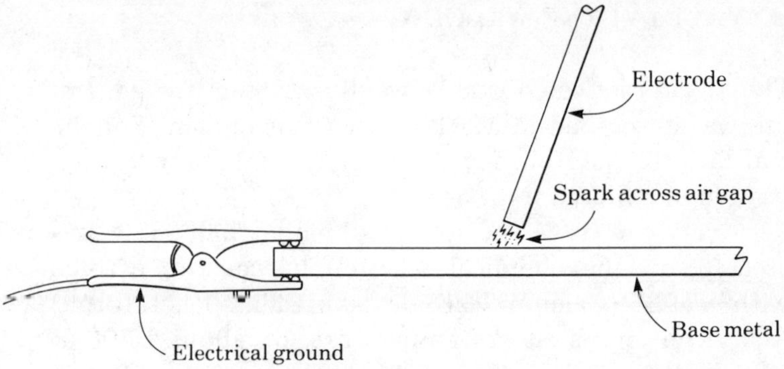

Figure 1-6. Principle of the arc welder. Electricity flows through the electrode, which is held a short distance (usually the distance equals the diameter of the rod) above the base metal to be welded. Electrical voltage (pressure) causes the electricity to jump the air gap (distance between the end of the electrode and the base metal). The resistance of the spark to the air gap generates very high temperatures that melt the base metal.

Arc welding uses an arc of electricity to melt metal. The electric arc is as old as the weather because it is the same thing as lightning, which leaps the air gap between stormy clouds and the ground. For practical purposes, we may call the arc "a flame of electric fire." Electricity is pushed through a wire (conductor) by pressure called *voltage*. An arc is created when the voltage is strong enough to force the electrical current to jump across an air gap. Heat is generated when the electrical current increases to overcome the resistance of the air gap. We can define resistance as something that is difficult for the electricity to flow through. The more resistance the electricity has to get through or across, the greater the amount of heat generated.

In arc welding the electricity flows through wires to a holder that grips an electrode. An electrode is a small rod of the same metal as that being welded. The electrode tip is then held above the work to be welded at a distance equal to the diameter of the rod. This distance is the *air gap*. In jumping this gap, the electric arc generates a temperature of 6000 to 9000° F. This temperature is more than sufficient to melt and fuse the base metal.

WELDING SYSTEMS AND MACHINES

Some welding machines and systems are based on variations of either arc or oxyacetylene processes, but others are completely different. Friction welding is one of the different systems. Two pieces of metal are rubbed together so fast that the edges melt and fuse.

The Metal-Arc Stud Welder is one machine. It is small and fits over a stud and welds the stud to the base metal through a variation of arc welding. The Seam Welder uses electrodes carried in wheels. The wheels roll thin metal between themselves. This creates a continuous weld seam by making a series of overlapping spot welds.

The Electron Beam Welder is still another unusual machine. The welding heat is created by a stream of electron beams. Since the work must be done in a vacuum chamber to prevent air from stopping the electron flow, electron beam welding is used only for metals that might react chemically with substances in the air.

Users of some of these welding machines cannot properly be called *welders*. They are really welding machine *operators*. It is not unusual for a very good welder-operator to go home and be unable to weld a pipe under the kitchen sink. Most of the big semiautomatic machines are for specialized purposes only, and training to operate them does not go much beyond the needs of the machine.

For this reason, the manual welder with his portable equipment is still the backbone of the welding industry. The welder is in demand, and there is no foreseeable machine to replace him.

A WELDER AT WORK

It is easier to understand a welder's job if we stop for a minute and watch a welder at work. Oxyacetylene and arc welding are the mainstays of the welding trade, so we shall begin with these two processes. There are many tricks to the welder's trade, but they are not hard to learn if you approach them step-by-step. But each step must be learned thoroughly and practiced before you go to the next one.

The welder we will watch will be making a simple *butt weld*. That is, he will be welding two pieces of metal end-to-end. He will use the oxyacetylene process on flat metal 1/8 in. thick. Watching a welder at work can be dangerous. The brilliant light can cause temporary discomfort to the eyes, and radiation and splattering metal can cause burns. You must adequately protect yourself.

In this brief overview we will only sketch what a welder does to provide a basic idea of the work so that

Figure 1-7. The oxyacetylene setup consists of an acetylene gas tank (left), regulators, hoses, and a torch (right). This portable setup is mounted on a hand truck for easy moveability.

more detailed explanations later will be easier to understand.

A portable oxyacetylene outfit consists of a tank of oxygen and a tank of acetylene gas. They are mounted side-by-side on a wheeled cart so they can be easily moved to the work site (Figure 1-7). Each tank has valves and regulators and a hose that runs from each regulator to the welding torch. The torch has valves so the operator can adjust the oxygen and acetylene to obtain the right flame for his work (Figure 1-8). The working end of the torch has interchangeable nozzle tips.

PREPARATION FOR OXYACETYLENE WELDING

The operator has two important jobs to do before he begins welding: He must identify the metal he is working with and clean the metal before heating it.

Identifying metal is not easy because many metals look alike to the eye. But there are tests that will identify the metal. The welder cleans the metal because welding is a melting process. Thus, any rust, dirt, or impurities on the metal will melt and be carried into the weld. This can produce a weak weld that may break when the welded article is put back in use.

BUTT WELDING

The metal is ready for welding after it has been cleaned. Very thin pieces of metal can be welded just by butting together the two pieces (Figure 1-9). Thicker metal must be beveled along the mating edges to enable the acetylene flame to penetrate all the way through the metal. Otherwise, we will not achieve full penetration. This beveling may be done by grinding, chiseling, or flame cutting with the oxyacetylene torch.

Beveling provides a V-shaped groove so that the welding flame can reach all the way to the bottom. Filler rod is melted along with the base metal to fill the groove left by the beveling.

After the metal is cleaned and beveled, the welder adjusts the flow of oxygen and acetylene through his torch. He then lights the torch with a "striker." This is a flint and steel scraper that creates sparks to ignite the

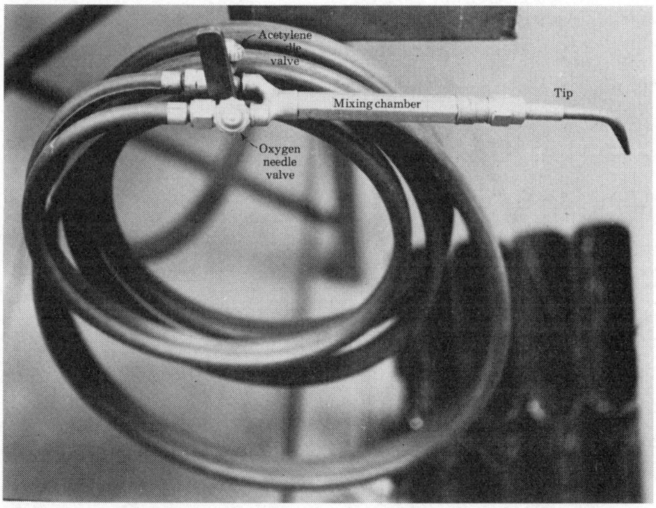

Figure 1-8. Oxygen and acetylene are mixed together in the oxy-acetylene welding torch.

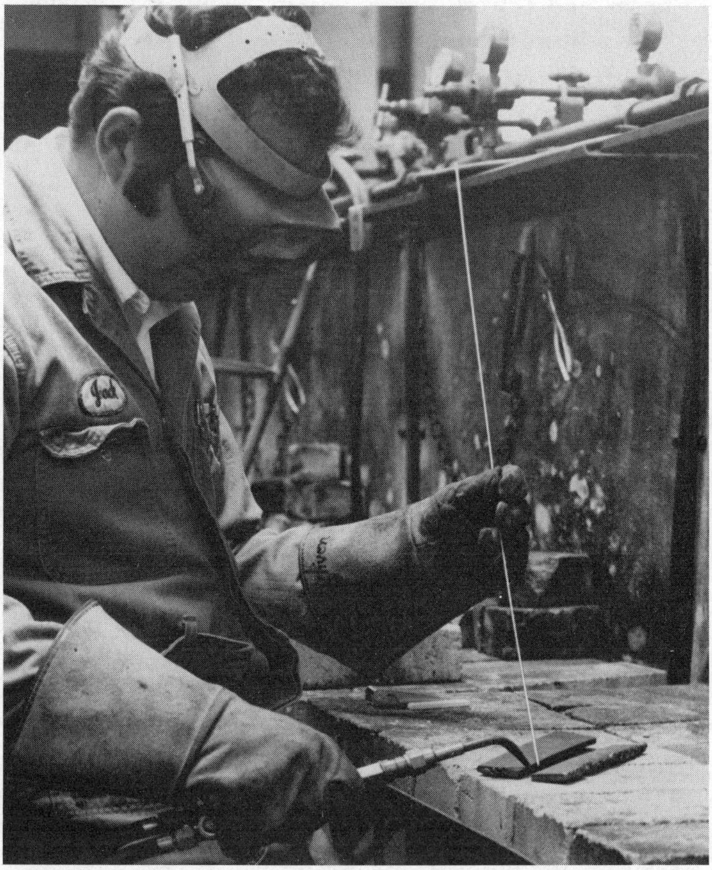

Figure 1-9. The welder makes a butt weld by heating the edges of two pieces of metal and adding filler rod to fuse together the two pieces.

torch. Matches are unsafe to use to light the torch. The flare of fire when the torch ignites could cause a disastrous burn even through heavy gloves.

PUDDLING AND FILLING

Next the welder pulls his heavily shaded glasses over his eyes and adjusts the torch to produce the correct flame. He holds the torch in one hand and the filler rod in the other, working in this manner:

1. The flame is used to melt a single spot on the base metal. This spot is called a *puddle*. The end of the filler rod is dipped into this molten puddle where it is melted by the puddle's heat. The rod is never melted by the flame. This supplies the extra metal needed to fill the beveled groove in the base metal.
2. The welder rotates the torch slightly to form overlapping ovals of molten metal. He continually raises and lowers the rod as extra metal is needed. This requires coordination.
3. The torch is advanced a short distance each time to create a continuous puddle. The resultant weld bead will have a "ripple" appearance. The bead is a line of hardened, continuous ripples formed by the progressive cooling of the puddles.
4. At the end of the bead, the torch is extinguished by turning off the fuel valve first and the oxygen next. This purges the hose and torch of flammable fuel. All valves are then retightened.
5. Finally, after the weld has cooled sufficiently, the welder inspects the bead for quality.

AN ARC WELDER AT WORK

The arc welding process uses an electrical machine (Figure 1-10) to provide the proper amperage and voltage for the job. Wires from this machine run to an *electrode holder*, commonly called a "stinger." This is an insulated clamp to hold the *electrode*. The electrode is a metal rod that serves as a current conductor for electricity and to take the place of the filler rod used in oxyacetylene welding (Figure 1-11).

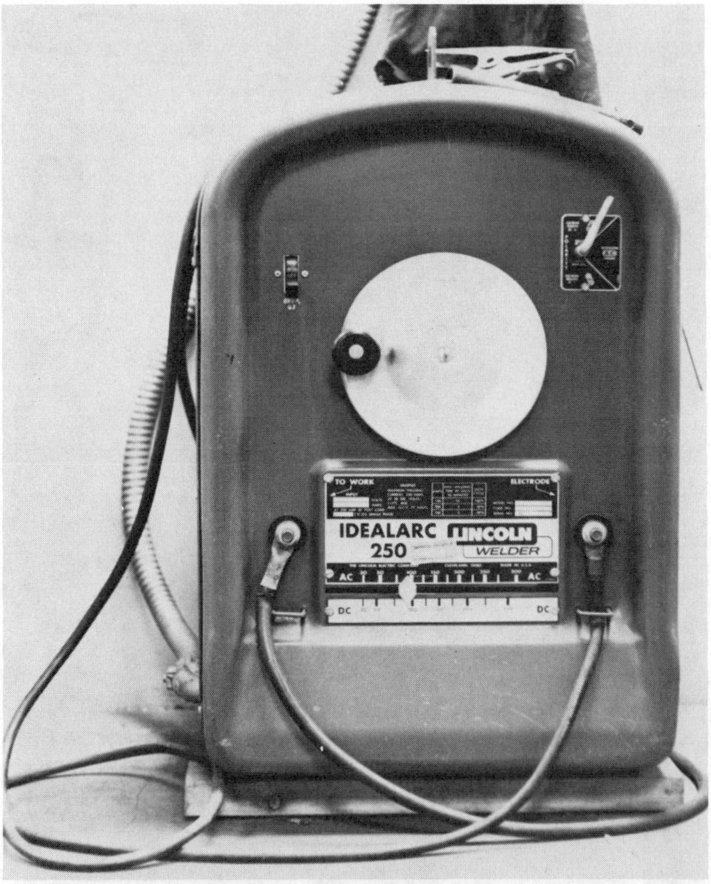

Figure 1-10. Welding machines provide the proper amperage and voltage for arc welding.

Figure 1-11. The electrode used in arc welding is gripped by an electrode holder.

Figure 1-12. Arc welding requires heavier protective clothing than oxyacetylene welding.

Electricity will flow only through a completed circuit. Therefore, a ground cable is attached to the base metal to provide the circuit for the electricity to flow back to the machine.

Arc welding produces more metal splatter and radiation than oxyacetylene welding. This requires the operator to wear heavier protective clothing (Figure 1-12). He also wears a helmet instead of goggles. He must also use a heavier shade of filter for eye protection.

STRIKING THE ARC

The arc welder starts his electric fire by striking an arc. He does this by turning on the current and either tapping the electrode against the base metal or scratching the base metal with a match-striking motion. This touch closes the circuit. Electricity flows from the machine and through the electrode, base metal, and ground cable back to the machine.

After the current creates the arc, the welder raises the electrode tip to a height approximately equal to the electrode's diameter, but not over 1/8 in. regardless of the diameter of the electrode. This creates the *arc gap*. The heat is created by air resistance as the current leaps this gap.

ARC WELDING PROCEDURE

1. The hot arc melts the base metal, forming the puddle of molten metal called the *arc crater*.
2. The force of the arc stream's penetration causes the molten metal to pile up at the back of the crater.
3. As the arc moves slowly down the seam, this piled-up metal cools into a series of ripple ridges, called "fish scales," forming the weld bead.
4. Flux, the chemical coating on the electrode, melts to form a gaseous shield about the arc crater. This shield prevents air from acting on the metal (oxidation). The flux also combines with impurities in the molten metal to create a *slag* that floats to the top of the puddle. The slag cools and crystallizes on top of the bead. It must be chipped off after the weld cools.

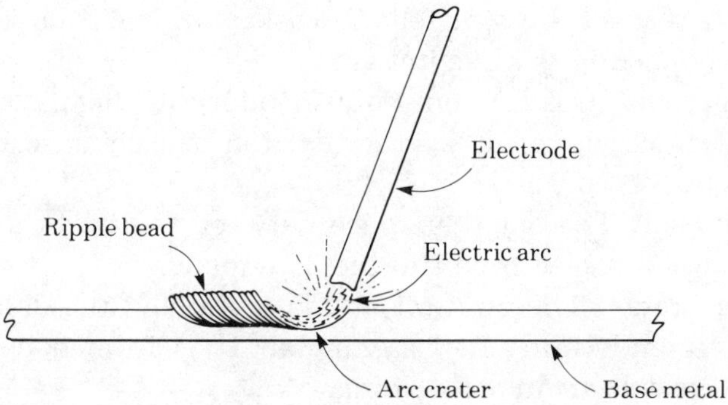

Figure 1-13. Principle of the arc welder. Electricity flowing through the electrode held by the welder creates an arc as it leaps the air gap between the end of the electrode and the base metal. The force of the arc digs an arc crater in the base metal, which has been melted by the heat of the arc. The arc pressure causes the molten metal from the arc crater and melting electrode to pile up at the back of the crater. These "ripples" make up the weld bead.

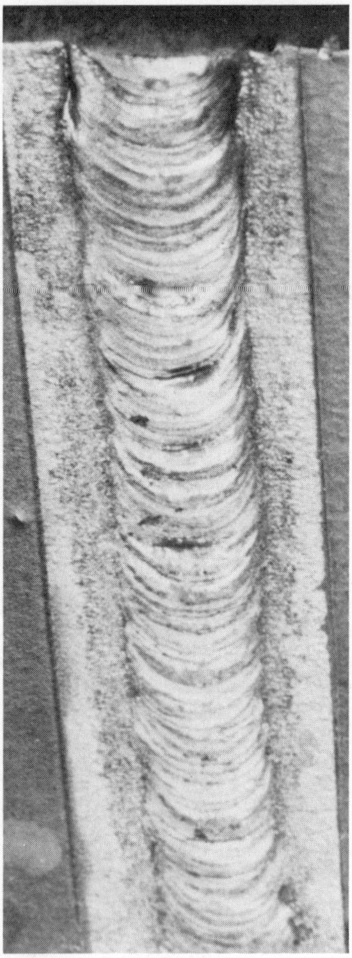

Figure 1-14. The force of the electric arc piles up molten metal in rings to create the rippled bead pattern characteristic of art welding.

MORE THINGS TO LEARN

This too-brief description of oxyacetylene and arc welding makes the job sound simple. It is simple—after a great number of things have been learned thoroughly step-by-step before you go on to the next portion. Thus you will gradually overcome your initial awkwardness and will soon be able to lay a bead with the best.

SUMMARY

1. Welding is one of the most widely used professions today.

2. The two most common welding processes are oxyacetylene welding and arc welding.

3. Oxyacetylene welding uses a mixture of acetylene gas and oxygen to produce a 6300° F flame. The flame is used to melt the edges of two pieces of metal and fuse them together. Filler rods, melted into the puddle, add required metal.

4. Arc welding uses an electric spark created by an air gap (which is also called an arc gap). This spark is directed to the area to be welded by an electrode in an electrode holder gripped by the operator.

5. The force of the arc creates an "arc crater." Metal piles up at the back of this crater and cools in ripples to create the arc weld bead.

NEW WORDS AND TERMS

Although the purpose of this book is to eliminate as many hard-to-remember technical terms as possible, every profession develops a language of its own. You must learn some of these words if you are to understand your teacher, boss or fellow workers. Some such words introduced in this chapter are

Acetylene Gas A poisonous, odorless, highly flammable gas made by the reaction of calcium carbide pellets to water.

Arc Gap The open air space between the tip of an arc welding electrode and the base metal.

Arc Crater The pool of molten metal and the depth it extends into the base metal. It is formed by the electric arc in arc welding.

Arc Welding A welding process that uses the principle of heat generated by electrical current leaping an air gap.

Arc Welder The equipment or "outfit" used in arc welding.

Base Metal The material that is being welded.

Bead The hardened line of fused metal that created the weld.

Blowpipe Slang for the oxyacetylene torch.

Butt Weld Two pieces of metal welded end to end.

Electrode The rod held in the electrode holder that serves as a conductor for the electricity. The melting tip also provides filler metal for the weld.

Electrode Holder A heavily insulated clamp that holds and transmits electricity to the electrode.

Filler Rod A small rod, used in oxyacetylene welding. It is melted into the puddle to provide extra metal needed to fill a weld.

Forge Welding Where metal is heated until it softens and then is hammered into shape. Blacksmithing is a common example.

Fusion Welding A welding process whereby metals are melted along their edges and fused together.

Oxyacetylene Welding A type of fusion welding that uses the flame of acetylene gas and oxygen for heat.

Puddle Molten pool of metal melted by an oxyacetylene flame.

Slag Impurities in metal that float to the top of the weld bead.

Striker A flint and steel that strikes sparks to light welding torches.

Thermit Welding A welding process that uses the reaction of aluminum and iron oxide to weld large pieces of iron.

Torch, Cutting An oxyacetylene torch with an extra outlet for pure oxygen, used in cutting metals.

Torch, Welding A mixing device equipped with a flame nozzle to burn welding gases and oxygen. The oxyacetylene torch is the most common of this type.

Welder One who welds. Often spelled weld<u>or</u>, to differentiate the person from the welder or welding machine.

PERSONAL SAFETY

The first rule a beginning welder must learn is: Take care of yourself. Welding is not considered a dangerous profession, but it can certainly become one if the welder is careless or violates the basic safety rules.

We are working with dangerous materials. The oxyacetylene gas used in welding is highly flammable. The oxygen used with it is stored in cylinders under 2,200 pounds (lb) of pressure per square inch. Knocking over an oxygen cylinder so that the gauge breaks off can turn the cylinder into a rocket.

In arc welding, we are working with high voltages of 220 to 440 V. Both the arc and the oxyacetylene flame are very hot. The temperature of an arc welding setup may range from 6000° to 9000° F. The heat generated by an oxyacetylene flame is from 5500° to 6300° F. Being burned by such heat is almost like touching the surface of the sun itself, which has a surface temperature of 11,000°F.

Even if we are careful about handling the arcs and flames and the high voltage, there is still the possibility of disastrous explosions from the materials we work on. A tank or pipe that once held combustible materials will usually retain a film of oil or fumes from gases. These can explode with devastating force when ignited by the extremely high temperatures of the welding arc or torch. There is also the possibility of toxic (poisonous) fumes caused by the heating of certain metals or alloys.

These and other dangers from improper use of welding equipment will be treated in detail as each individual piece of equipment is introduced and its use

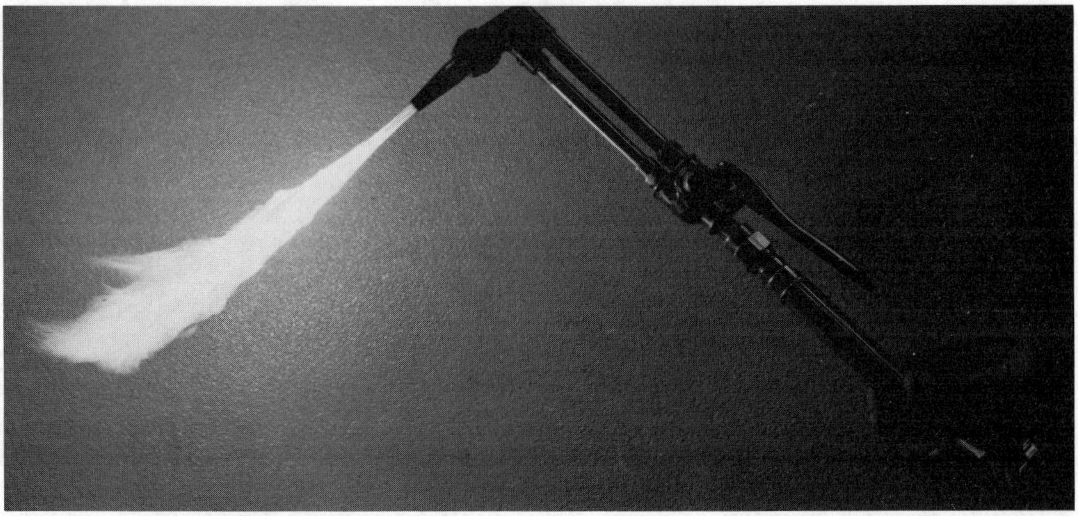

Figure 2-1. An oxyacetylene cutting torch generates heat ranging from 5500 to 6300°F. It must be handled with extreme care to avoid serious burns.

explained. This will make the information available exactly at the time you need it. In this chapter we are not concerned with the equipment but with your personal safety.

A SAFE PLACE TO WORK

Your first consideration is a safe place to work. Ideally, a welder should work in a special welding booth in a permanent location. This booth would include a work table, jigs and clamps to hold different types of pieces, and easily accessible working equipment. The floor should be either concrete or some other nonflammable material. The walls should be fireproof, and there should be curtains to protect the eyesight of others in the room. There should be a metal canopy over the top of the booth, equipped with an exhaust fan to draw away smoke and toxic fumes created by the welding process. In the past, curtains of an opaque fireproof material were used around the welding booth to protect the eyes of others working in the area. But recently a transparent curtain material has been developed that contains radiation-filtering material to accomplish the same results.

In all cases, keep grease, oil, and any flammable materials out of the welding booth. If it is necessary to lubricate any valves or equipment, use glycerine or

Figure 2-2. Special curtains containing radiation-filtering materials have been developed to replace the heavy opaque curtains formerly used. Even so, as this sign warns, you should not stop and watch a welder at work.

liquid soap. If the booth is used for arc welding, take care to ensure there is no dampness on the floor to prevent any short circuits or electrical grounds.

SAFETY OUTSIDE THE SHOP

Safety is not difficult to ensure in the shop because conditions can be controlled. It is a different matter when welding must be done with portable or home equipment outside a professional welding shop.

When working away from the shop, the first thing to watch out for is an explosive atmosphere. This can be anywhere, even in the middle of a street. A bad incident some years ago involved a car smashup near a welding shop. The car body was crushed so badly that the driver was trapped inside. A welder was called to cut the crumpled metal from around the victim. Unfortunately, the gas line had broken. Either raw gasoline or gas vapors were ignited by the welding torch being used to cut through the crushed metal. The trapped man was burned to death, and the welder was badly injured.

Similar explosive situations can arise on the job or in a handyman's garage. Greasy engines, containers that held combustible materials, or even close quarters where explosive vapors may collect are all sources of potential danger. Care must also be taken to ensure that the immediate area around the welder is cleared of debris, trash, and loose paper. Home welders often overlook these basic precautions.

PROTECTIVE CLOTHING

The flame from the oxyacetylene torch, the arc from the arc welder, and the molten pools of metal in both processes give off tremendous heat, glare, and radiation. The operator's eyes, hands, and body must be protected.

Radiation produced by arc welding is stronger than that of oxyacetylene. The arc flame is particularly strong in ultraviolet—the ray that produces sunburn—and infrared heat rays. These radiations are powerful enough to burn through light clothing. A good cotton material is recommended, or coveralls should be worn

Figure 2-3. Bits of flying metal—red hot "cherries" or "berries", as welders call them—can drop into pant cuffs, and open pockets, and even down your collar. This welder's leather jacket has fasteners on the collar to ensure a tight fit that will prevent sparks from going down his neck.

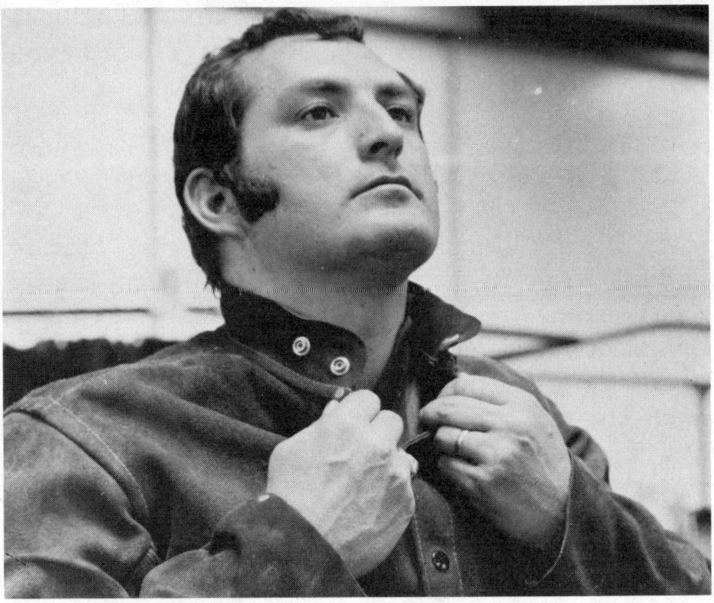

over regular clothing. The coveralls should be treated with a fire-retardant chemical.

Flying particles of hot metal can easily lodge in trouser cuffs, pockets, and even shoe tops, so it is a good safety practice to avoid cuffs on trousers, open pockets, and pleats. Pants should be long enough to cover shoe tops. Boots are preferable.

Figure 2-4. A welder's hands are particularly vulnerable to burns and should be protected by leather or asbestos (no cloth) gauntlet-type gloves. Note that this welder is wearing a leather jacket and leather pants for additional protection.

Gloves

Your hands, working so close to flame, arc, and molten metal, are especially vulnerable to radiation burns and splattered metal. They should be protected by leather or asbestos gauntlet-type gloves (Figure 2-4) that cover the wrists. The gloves should not be too thin. When doing overhead welding, where there is greater splatter and droppings, sleeve protectors and shoulder coverings should be used.

PROTECTING YOUR EYES AND FACE

Special goggles are used by the oxyacetylene welder to protect his eyes. Arc welding produces more radiation and metal splatter than does oxyacetylene, and goggles

Figure 2-5. Special goggles with dark filter lenses are worn for oxyacetylene welding to protect the welder's eyes. This pair has ventilating grids in the side of the eyepieces.

are inadequate protection. Arc welders use helmets that cover the entire face.

It is important to protect one's eyes from his own work, but it is equally important to protect yourself from the flames and arc flare of other welders working near you. It is considered dangerous to watch an arc with the naked eyes at any distance closer than 200 feet (ft).

Goggles worn for oxyacetylene welding fit tightly around the eyes. They cut out all light except that which comes through the lenses. This light is heavily filtered to remove harmful radiation and glare (Figure 2-6).

Figure 2-6. The unigoggle type has one eyepiece to cover both eyes. It also features an adjustable head band. The eyeshade pivots so it can be pushed up to permit the welder to inspect his work after the torch has been extinguished. *This type of goggle is superior to the goggles shown in Figure 2-5.*

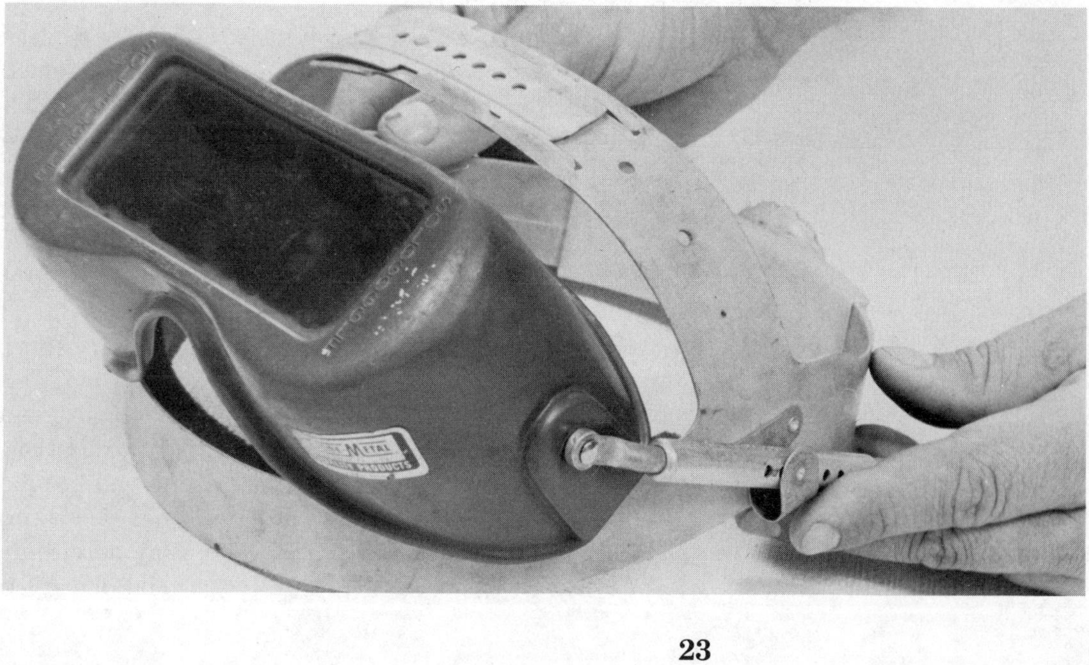

Figure 2-7. The dark filter (top) is protected by a clear glass or plastic shield (bottom). The splatter on the clear shield shows how the protective glass or plastic becomes fouled by splatter and pitting. Plastic is preferred to glass because glass may shatter.

Eye Filters

The filters used in both oxyacetylene goggles and arc helmets are expensive, so they are protected by sheets of plain glass or plastic in front of the filter glass. Plastic is recommended because it will not shatter as glass can. This protective cover is replaced when it becomes splattered or pitted. In addition to protecting the eyes, the filter cuts down light brilliance so the welder can clearly see the crater and puddle to guide him in his work.

The depth of filter colors range from number 1, the lightest shade to number 12, the darkest. The less brilliant oxyacetylene welding usually requires a number 4 or 5 filter. Arc welding, with its greater radiation, uses a number 10 or possibly a number 11 or 12. No hard rule can be drawn because each individual's eyes may require different protection. If you see white spots before your eyes after welding, your filter is too light.

Arc Welding Helmets

The arc welding helmet is usually made of fiber board or fiber glass. It fits over the skull cap that welders should—but sometimes do not—wear. The helmet encloses the entire face and neck, half of the top of the head, and both sides of the face.

A rectangular opening level with the welder's eyes is covered with shaded filter glass and an outside layer of plain glass to protect the shaded filter from splatter. The filters can be changed to provide the proper density for different jobs.

Perhaps sometime you have accidentally looked at the sun or at an exceptionally bright light. When you looked away you continued to see bright, white spots before your eyes. If you get this effect after welding, it is a sign that you are using too light a shade of filter in your goggles or helmet.

Filter density. The density of the shaded filters, as mentioned earlier, ranges from 1 to 12, with number 5 being the average density for oxyacetylene and number 10 the average density for arc welding. It is not necessarily the brightness of the glare that determines the density of the shaded filter chosen. The density of

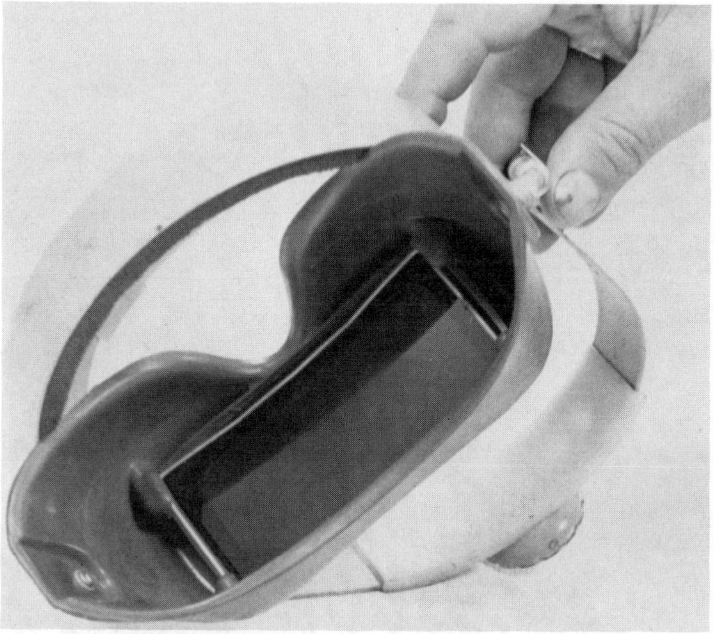

Figure 2-8. The shaded glass filter is held in this type of goggles by a spring clip, which permits the filter to be easily exchanged for a darker or lighter shade when necessary. Oxyacetylene welding generally requires a number 5 shade.

Figure 2-9. This is one of the less expensive types of welder's helmets. It is made of fiberboard construction. Safety glasses must be worn with this type of helmet.

Figure 2-10. On this molded fiberglass arc welding helmet the filter cover swings up to permit ease in changing the shaded filters. The density of the filter must be sufficient to remove the harmful ultraviolet and infrared rays. The flip-up lens has a clear glass safety lense built in behind it to allow the welder to see his work after the weld is made.

the filter must be sufficient to remove the invisible ultraviolet and infrared rays. The heavier filters (the higher numbered) will remove from 99.5 to 100 percent of these rays. Infrared rays (heat) in excess amounts cause internal injury to the eyes. Ultraviolet rays can cause very painful conditions in the back of the eyeball. This pain usually appears about 8 hours (hr) after the eyes have received an excess exposure.

Face Shields

The welding helmet is intended to protect the welder's face and eyes from metal splatter and harmful rays during the welding process. It is not adequate protection for grinding joints, chipping slag from beads, or using a wire brush to clean metal before welding. The dark filter would prevent you from seeing what you were doing.

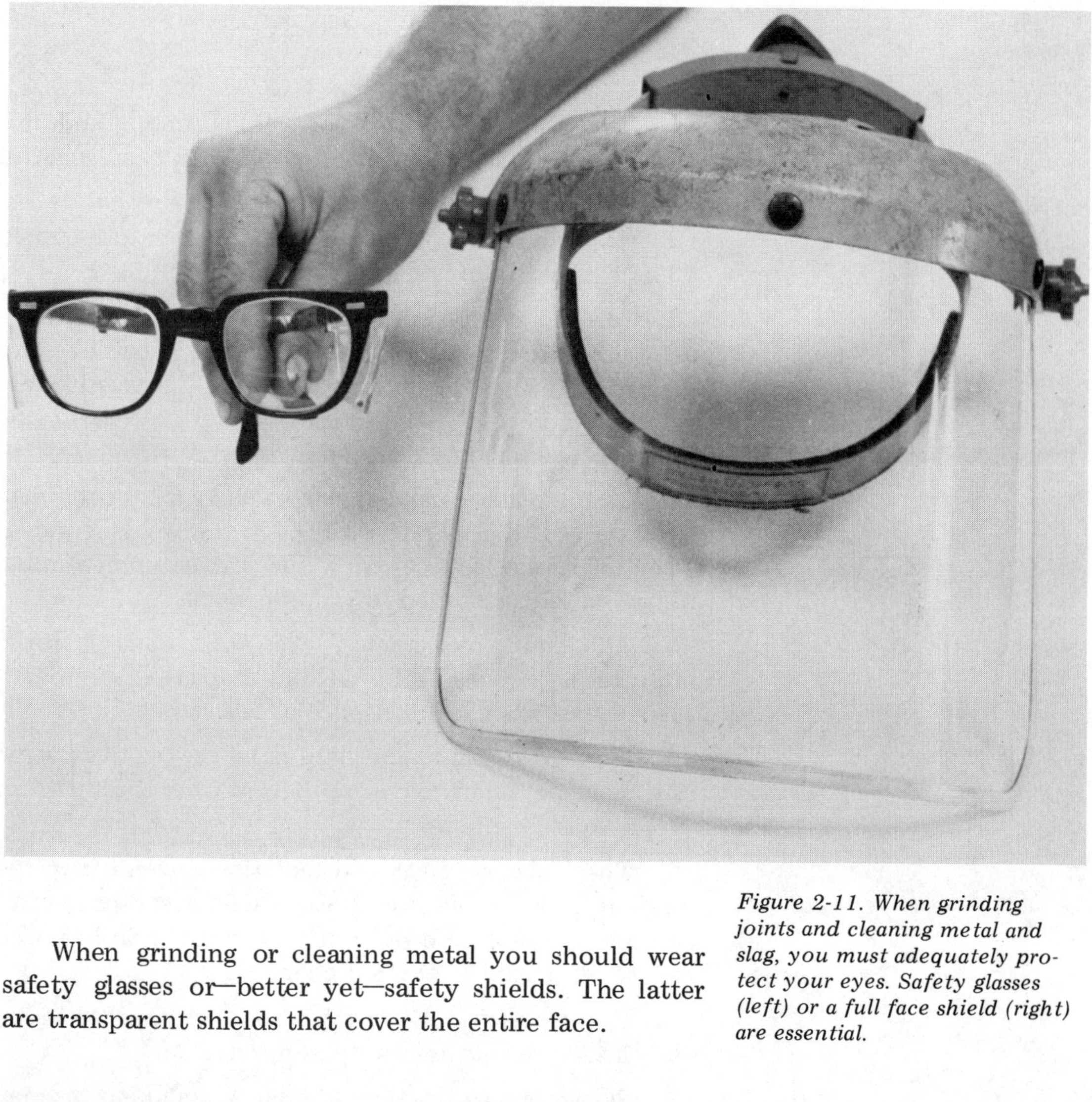

Figure 2-11. When grinding joints and cleaning metal and slag, you must adequately protect your eyes. Safety glasses (left) or a full face shield (right) are essential.

When grinding or cleaning metal you should wear safety glasses or—better yet—safety shields. The latter are transparent shields that cover the entire face.

FIRST AID

It cannot be overemphasized that burns and eye injuries are very serious matters. They must be given immediate treatment. Light burns are treated like sunburn or other minor burns, but more serious second- and third-degree burns must be treated by a doctor as soon as possible. No burn or injury of any kind should be neglected.

Every welding shop should have a first aid kit and each person should be trained in Red Cross first aid procedures. However, first aid is an emergency measure. It is not a substitute for prompt medical attention for the injured person.

SUMMARY

1. The high temperatures of welding flames and arcs emit glare, heat and radiation. These can cause painful burns and permanent damage to the eyesight.

2. Oxyacetylene torches burn at temperatures up to 6300°F. Welding arcs range from 6000 to 9000°F.

3. Personal safety begins with a safe working environment. In the shop a welding booth should be lined and floored with fireproof material. It should have adequate ventilation because the welding process releases toxic fumes from some metals.

4. Care must be taken to ensure there is no dampness on the floor, the work bench, or in the welder's gloves when using an electrical arc welder. All equipment must be adequately grounded to prevent shocks.

5. All flammable greases, oils, waste, and material must be kept out of the welding booth. Glycerine or soap is used when lubrication is necessary.

6. When working in the field or in the home extreme care must be taken to ensure that no work is done in an explosive atmosphere.

7. The excessive heat and radiations given off in the welding process require that welders wear protective clothing, preferably heavy cotton material that has been treated with flame-retardant chemicals.

8. Leather gloves, high-top shoes, cuffless trousers, and closed pockets are safety requirements.

9. Sleeve protectors, shoulder guards, and aprons may be required for specialized jobs, such as difficult position overhead welds where there is danger of drip.

10. The eyes are especially vulnerable. They must be protected by goggles and helmets fitted with heavily shaded filters.

11. Goggles with number 5 filters are usually adequate for oxyacetylene welding. Arc welding, with its greater splatter and radiation, requires a helmet that protects the entire face and neck. A number 10 to density filter is required for arc welding.

12. Goggles with safety glass or a face shiel worn when grinding, cleaning, or chipping m

NEW WORDS AND TERMS

Filter lens Dark glass, usually green or amber, used in welding goggles and helmets.

Helmet A hood for the head, equipped with filters for eye protection.

Infrared Rays Invisible rays beyond the red end of the visible spectrum. They have a deeply penetrating effect that can permanently damage the eyes.

Splatter, Metal Bits of molten metal thrown off by the welding process. Splatter, caused by the force of the arc, is greater in arc welding than in gas welding.

Ultraviolet Rays Invisible rays of radiation lying beyond the violet end of the visible light spectrum. They produce sunburn and can damage the back of the eyeballs.

SOMETHING ABOUT METALS

Metallurgy is the science of metals. It begins with ore buried in the ground and continues through the use of refined metal in industry. It is possible for one to become a passable welder without knowing much about metallurgy, but that would be doing it the hard way.

Metallurgy is a complex science, and many people find it difficult to understand. Fortunately, the part we need to know to be good welders is comparatively easy to learn.

There are two major reasons why we have to know something about the science of metals to make good welds. First, we must be able to identify the type of metal we are asked to weld—different metals require different processes. Second, metals change their properties and behave differently when heated than they do when cold. In welding we are working with heat on metal. Since various metals react differently to heat, we must understand exactly what heat does to particular metals.

WHAT HAPPENS WHEN METAL MELTS

Steel is a mixture of pure iron and varying degrees of carbon. If we inspect a piece of low carbon steel under a magnifying glass, we can see that the metal looks grainy. These grains are crystals of metal. All metals solidify in the form of crystals.

The larger these grains, the softer the metal. The smaller the grains, the harder the metal. Heat causes a change in the size of the grains.

As the steel is heated, the first thing we notice is

that its color changes. First there is a faint red that turns to a blood red. The metal then goes through a succession of color changes from cherry red to salmon orange, lemon yellow, light yellow, and then white hot before melting occurs. This color change relates only to iron and steel. Some metals, like aluminum, do not change color up to the point of melting.

Critical Temperature

Changes in color and expansion caused by heat are the outward signs of what the rising temperature is doing to our piece of low carbon steel. Inside the metal an unusual change is taking place. From room temperature to just before 1325°F there is no change in the steel's grain size. The color, however, has changed to a medium cherry red.

The 1325°F temperature is known as the *lower critical point*. It is the beginning of the *critical range* for this metal. The critical range extends from 1325° to 1550°F.

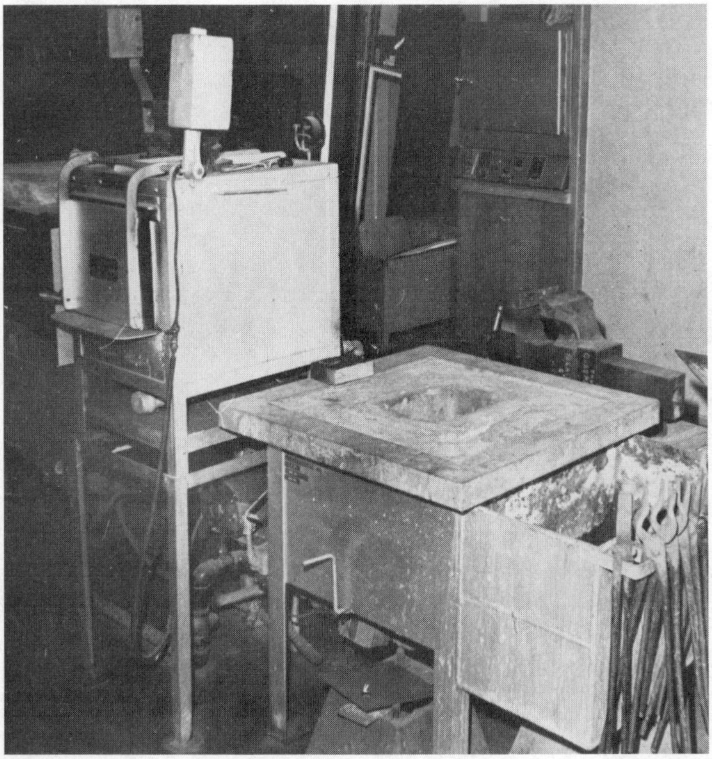

Figure 3-1. Another very useful combination when working with metals is a closed furnace and an open furnace. Inside either type of furnace, temperatures can be controlled.

How the Grain Changes

At 1325° F the normal cold-state grain size of this piece of low carbon steel begins to shrink. It grows progressively smaller as the temperature increases to 1550° F. Beyond this temperature, *the upper critical point*, the grain size begins to increase again. This increase continues until the steel melts at 2786° F. The grain disappears when the metal melts. The change at 1325° F is caused by the the atoms of the metal beginning to rearrange themselves into a new structure.

Melting Temperatures of Metals

Different metals melt at different temperatures. Here are some melting points of metals welders come in

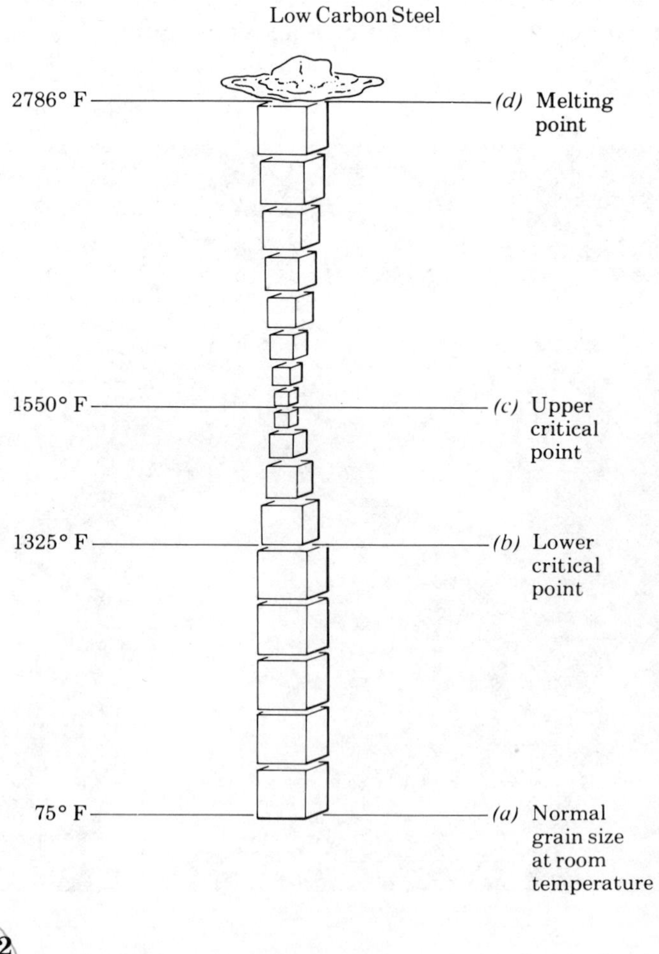

Figure 3-2. This graph depicts the action of low carbon (mild) steel grains when heated from temperature to a molten state. The normal grain size (a) does not change until the temperature reaches the lower critical point (b) at 1325°F. The grains then decrease in size until they reach the upper critical point (c) at 1550° F. They then increase again in size up to the point of melting at 2786° F. Above the melting point the grains lose their cube shape.

contact with, either as pure metals or as alloys (mixtures) with other metals.

Metal	*Temperature, °F*	*Temperature, °C*	*Chemical Symbol Of the Metal*
Aluminum	1212	659	Al
Beryllium	2332	1277	Be
Boron	4200	2316	B
Carbon	6700	3828	C
Cobalt	2696	1480	Co
Copper	1981	1083	Cu
Chromium	1520		Cr
Iron	2786	1530	Fe
Lead	621	327	Pb
Manganese	2300	1260	Mg
Molybdenum	4730	2610	Mo
Nickel	2646	1420	Ni
Silicon	2588	1420	Si
Silver	1761	960	Ag
Tin	450	232	Sn
Titanium	3035	1668	Ti
Tungsten	6170	3410	W
Vanadium	3450	1900	V
Zinc	787	419	Zn
Zirconium	3366	1852	Zr

WHEN METAL COOLS

Metal cooling from the molten state goes through the melting process in reverse. The crystallized grains begin to form in the liquid metal when their temperature drops below the melting point. These grains will be large but will decrease in size as the metal cools. They will reach their smallest size in the critical range and then increase to their normal cold-state size.

This happens only if the metal is permitted to cool normally. If the steel is cooled more quickly, the grains will remain small. If the metal is cooled very slowly, grain size will be enlarged.

Since fine grain steels are harder and large grain steels softer, we can use the cooling rate to make some metals harder or softer by freezing the grain pattern at a particular point. We do this by processes called *quenching* and *annealing*.

Quenching

Quenching means to cool quickly. We can do this by suddenly plunging the hot steel into water or a mixture of oil and water or other cooling substances. The sudden drop in temperature "freezes" the grain size at the point it attained when plunged into the quenching bath.

Suppose we have a piece of medium carbon steel and we want to make it harder. This requires reducing the size of the grain structure from large crystals to small crystals. We do this by heating the metal to the critical range where the grain size has diminished to the point we wish. This would be about 1550° F. The color would be a bright red. At this point we remove the steel from the heat and plunge it into the quenching bath. This cools the metal suddenly, solidifying the grain structure at its smallest and hardest size.

Figure 3-3. This oven is useful for many metal projects but is most useful as an annealing oven for small objects.

This process applies to medium and high carbon steel. Low carbon (mild) steel does not have enough carbon to harden by the quenching method. Low carbon steel can only be case or surface hardened.

This grain change is permanent only as long as the metal stays below 1325°F. If the metal is heated above this temperature, which would be in the critical range, and allowed to cool naturally, the grain pattern would return to its original cold-state large grain size.

Annealing

Reducing the grain size of medium and high carbon steel increases their hardness. But hardness is not always desirable. Extra-hard metal is hard to work and difficult to weld. So in some cases we may find it necessary to soften metal to make it more easily workable or weldable. This is done by *annealing*. Annealing is the opposite of quenching. It means to cool the metal slowly at a controlled rate. This may be done in an annealing oven or by burying the metal in lime or sand to slow down the cooling rate. This slow cooling results in large grain crystallization.

Effect of Too Much Heat

The effect of heat on grain size is important to a welder. Our objective in welding is to maintain the original grain as nearly as possible in our base metal. If we use too much heat in welding, the extra heat may spread beyond the weld and change grain size in the adjoining metal. This could make this area softer, harder, or more brittle than the rest of the base metal. This change could cause cracks, warpage, or stresses that could lead to later failures.

HIGH CARBON STEEL'S CRITICAL POINT

We have used low carbon steel as our example. The amount of carbon determines the ultimate hardness of steel. Although low carbon steel can be heat-treated to reduce grain size and increase its hardness, it can never be made as hard as high carbon steel.

When high carbon steel is heated to the melting point, it acts the same as low carbon steel with one major difference. When low carbon steel reaches its lower critical point (1325°F), its grains begin to decrease slowly in size as the temperature increases to 1550°F. The atoms of high carbon steel also are rearranged at the critical point of 1325°F, but the rearrangement of high carbon steel atoms is sudden. It has no critical range as does low carbon steel. The grain size diminishes to its smallest point at this temperature, then it begins to enlarge as the temperature is increased to the melting point.

ALLOYS

In welding we rarely, if ever, work with pure element metal. Instead, it will probably be an alloy. An alloy is a mixture of a base metal with varying amounts of other metals. Small portions of alloying metals will change the characteristics of the base metal. We noted this earlier in the way an increased amount of carbon makes steel harder. Alloying metals can make a base metal tougher, elastic, bendable (ductile) or—as in stainless steel—less likely to rust.

An alloy metal will have different welding characteristics. In some cases the addition of these metals will make an otherwise weldable base metal unweldable. For example, low carbon steels are easily welded, but the addition of extra carbon to produce high carbon steels increases the difficulty of getting a good weld. High carbon steel melts under the flame or arc, but the puddle boils and cools to an imperfect weld that lacks strength.

BRAZING

In some cases different electrodes can be used in arc welding to weld difficult metals, but alloys that cannot be readily welded by fusion (melting) can often be joined by *brazing*. Brazing can be described as welding without melting the base metal. The filler rod used in brazing has a lower melting point that the base metal. The filler melts around the broken points, encasing them. A good brazing job may be as strong as a fusion weld. Brazing is also employed when high heat might upset the heat treatment of a base metal since the broken parts can be brazed at a temperature below the melting point of the base metal.

METAL TYPES

Metals are divided into two basic classes. One is *ferrous* metals, those that contain an iron base. The other is *nonferrous* metals, those that do not contain iron. Aluminum, copper, tin, lead, zinc, and the like are examples of nonferrous metals.

Iron

Iron is mined as an ore and processed in giant furnaces to reduce its impurities. The molten iron is poured into molds called "pigs," giving the name *pig iron* to the resulting product.

Pig iron is remelted with alloying amounts of carbon, silicon, manganese, sulfur, and phosphorus to make *gray cast iron*. This alloy is permitted to cool slowly in its mold. The slow cooling produces large crystals (grains) and causes the carbon to separate from the iron. This free carbon in the metal produces the characteristic gray color of this iron.

If the cast iron is cooled more rapidly than is done in making gray cast iron, there is less separation of carbon from the iron. This produces a harder, more brittle metal called *white cast iron*. White iron has a high degree of compressive strength and is used, among other things, for railroad train wheels. It can support heavy weights.

A softer iron for casting machine parts, called *malleable iron*, is made by softening white cast iron. This is done by heating the white cast iron to 1600°F for 40 to 60 hr and then cooling it slowly in the annealing oven to room temperature.

Later, if the malleable iron is heated by a welding torch above the critical temperature of 1325°F, the malleable iron in the areas subjected to this heat will change back into white cast iron.

Figure 3-4. Cast iron will sometimes show the marks of the sand mold used in its manufacture.

Still another type of iron is *wrought iron*. It is made of pig iron, iron oxide, and silica. Silica is the same material as sand. Iron oxide is iron and oxygen, or plain rust. Silica creates a reaction with the molten iron that produces a fibrous grain in the finished metal. Wrought iron has excellent rust-resistant properties.

Carbon Steel

Carbon steel—steel in which the only alloying element is carbon—is divided into four classifications:

1. **Low Carbon** This has a carbon content of 0.05 to 0.30 percent. It is harder and tougher than pure iron and more *ductile* than higher carbon steels. A ductile metal is one that can stretch, bend, or twist without breaking. Low carbon steel is also known as mild steel. It is easily worked by forming and machining and is easily welded. It cannot be heat treated, but it can be case or surface hardened.
2. **Medium Carbon** This is a carbon range from 0.30 to 0.45 percent. The higher percentage of carbon makes it harder and more difficult to work than mild steel. It is more difficult to weld and requires special arc welding electrodes. It can be heat treated.
3. **High Carbon** The carbon content here ranges from 0.45 to 0.75 percent. Steels up to 0.65 percent

carbon content can be successfully welded by using special electrodes and techniques. This steel is often used for top grade and is sometimes referred to as "tool steel."

4. **Extra High Carbon** This steel has a carbon content ranging from 0.65 to 1.7. It is very difficult to weld and requires special electrodes. It can be successfully brazed.

Alloy Steels

All carbon steels are alloys, of course, but when welders speak of "alloy steels," they mean metals alloyed with additional elements other than carbon. These elements are added to the basic carbon steel to create desirable characteristics.

Alloy steels are known by the name of the major alloying elements added to them. Thus we have chrome steel (chromium), manganese steel, molybdenum steel, nickel steel, tungsten steel, and cobalt steel. Some of these may be used in combination. Then there are chromium-high carbon steel, nickel-chromium-molybdenum steel, and others. Alloy steels are generally weldable without difficulty.

IDENTIFYING METALS

The different ways that metals react to welding heat make it absolutely necessary to know how to identify each metal. This is not easy since many of the metals look alike. But unless you know which metal you are working with, you will not know how much heat to apply, which filler rod to use, how the rate of cooling should be controlled, and the correct welding process or technique to use.

Several standard tests have been devised to aid welders in identifying metals. Some of these require experience beyond the ability of a beginner. The best way to learn is to take a piece of metal that has been identified by an experienced welder and subject it to all applicable tests. In this way you can observe the reaction of the metal to each test.

No one test is fully adequate because sometimes two metals will give the same reaction to a test. In such

Figure 3-5. The Rockwell hardness tester. The degrees of hardness can be used in distinguishing one type of carbon steel from another.

cases, you may need to perform two or more different tests to properly identify the sample.

Spark Test

The spark test is for ferrous metals, but this is not a disadvantage. The majority of metals a beginner works with will be ferrous.

The spark test is done by *lightly* touching a piece of metal to the revolving wheel of a power grinder. Ferrous metals will give off a stream of sparks. These sparks differ in length, shape, color, and "explosions." The explosions mean that a long spark may burst into a tiny shower of individual sparks, shooting off smaller sparks at an angle.

Figures 3-6 and 3-7 give some idea of how these different sparks look, but they can be understood only

Figure 3-6. The spark test is done by lightly touching the metal sample to a power grinder and observing the type, color, and length of the resulting sparks.

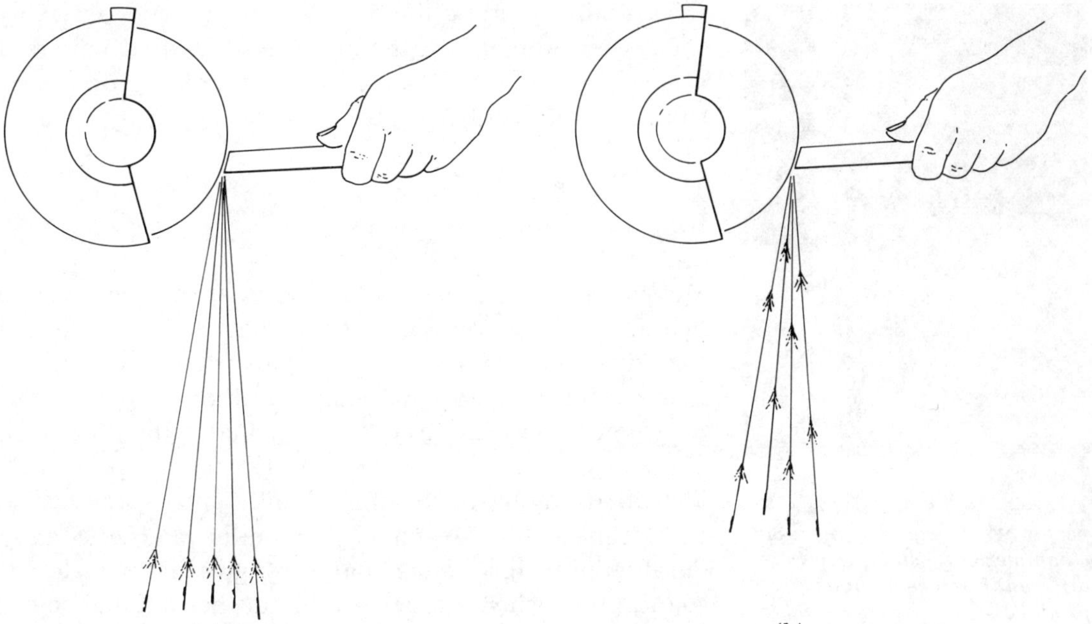

(a) (b)

Figure 3-7(a). Wrought iron. Shafts: long and ending in appendages; some forks near end. Color: yellow at top and white at bottom of bursts. Volume: large. (b) Low carbon steel. Shafts: long, but shorter than wrought iron. Ends in appendages with more forks than wrought iron. Color: white. Volume: moderately large.

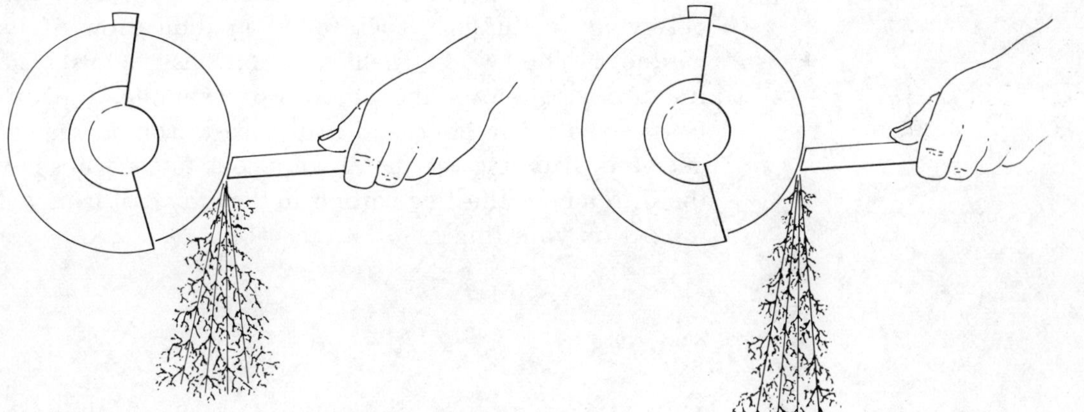

(c) (d)

Figure 3-7(c). Gray cast iron. Shafts: shorter than low carbon steel, with a large number of small and repeating bursts. Bursts differ from forks in that they branch out from the main fork with additional small forks. Color: red in the area near the grindstone, turning to straw yellow at the end of the streamers. Volume: small. Note: Gray cast iron and white cast iron look very much alike to the untrained eye when using the spark test. White cast iron has less volume and the bursts are finer than gray cast iron. (d) High carbon steel. Shafts: medium length, with many repeating bursts. Bursts increase as the carbon content increases in iron and steel. Color: white all the way. Volume: large.

Figure 3-8. The ring test is performed by striking metal with a hammer and listening to the different pitch each metal produces.

by actually grinding the different metals and observing the action. Remember, the metal must be lightly pressed against the grinder wheel. If you create too many sparks, it will be difficult to tell what you really have.

Ring Test

Metals produce sounds when they are struck with a hammer. The pitch of these sounds will be different for different metals. An experienced welder can identify many metals and alloys by the sound they give off when struck. Sounds cannot adequately be described. You can learn this test only by taking previously identified metals, striking them, and listening to the "ring." For example, aluminum and the alloy duraluminum look alike, but they give off a different sound pitch when struck. Aluminum has a duller sound than duraluminum.

Fracture Test

The fracture test requires breaking a piece of metal and observing the grain structure as an indication of its hardness. This works well with ferrous metals. The fracture also shows the color. For example, the difference between gray and white cast iron is readily detected. Rubbing the metal with your finger will show the difference—the free carbon in the gray cast iron will rub off on your finger.

Chip Test

The chip test identifies metal by its hardness. The test metals are cut by a chisel. Wrought iron cuts easily. The chips are large and the chipped surfaces smooth. Cast iron chips break into small pieces. Low carbon steel also cuts easily. The chips come off in long slivers and do not break easily. High carbon steel is hard to cut.

A file can also be used to distinguish between soft, low carbon steel and hard, high carbon steel. The high carbon steel is difficult to file. The file will just slip or slide on the metal surface.

Oxyacetylene Test

Tests can tell a welder what metal he is working with, but they cannot tell him if the metal can be welded satisfactorily. We may be told that a certain metal can be welded with a certain type of filler rod, but when we try it, we find the weld is unsatisfactory.

The reason is that impurities have gotten into the metal during the smelting process. These impurities can act as alloying agents themselves, changing the welding characteristics of the metal. Therefore, the final and most important test is whether the metal will weld satisfactorily. So we melt a puddle of steel or iron with the torch and see what happens. Watching the puddle through our welder's goggles, we look to see if the surface tension of the puddle is good. When the puddle hardens, the surface should be shiny and even. If it is dull, broken by small pits (indicating boiling during the molten stage), colored, or rough, then the weld is weak.

This metal is unsatisfactory for welding by common processes. There are processes today that can weld all metals; however, very exotic and expensive machinery must be used to weld metals which react to conventional oxyacetylene, gas-shielded arc, or heliarc processes.

IDENTIFYING NONFERROUS METALS

Nonferrous metals, with the exception of nickel, do not give off sparks (nickel sparks are meager themselves), so the spark test cannot be used to identify nonferrous metals. The spark test is used for short durations to show if a metal is ferrous or nonferrous.

Another way to distinguish between ferrous and nonferrous metals is by magnetism. Ferrous metals, with the exception of some stainless steels, are magnetic. It is necessary only to touch a magnet to the metal to distinguish between ferrous and nonferrous materials. Ferrous metals lose their magnetism when heated above a certain temperature, but all tests are made at room temperature.

Some nonferrous metals can be identified by their color. Copper, for example, has a very distinctive color that we are all familiar with because of the pennies we carry in our pocket.

Brass, an alloy of copper and zinc, has an opaque yellow color more golden in appearance than the reddish copper. Bronze, an alloy of copper and tin, has a deeper, more reddish appearance than brass.

Aluminum is distinguished by its light weight. It can be identified from its alloys by chemical tests. Pure aluminum stays bright in a 25 percent solution of sodium hydroxide (lye) and water. Aluminum alloys that contain copper will darken in this solution.

Specific Gravity Tests

When there is difficulty distinguishing between metals with closely matching characteristics, the specific gravity test may be used. Specific gravity is a comparison of the density of a substance with the density of water.

Archimedes, the Greek mathematician, discovered the principle. King Hiero of Syracuse suspected that his solid gold crown had been alloyed with silver. He asked Archimedes to prove if the goldsmith had cheated. The mathematician observed one day that when he got in the bathtub his body displaced some of the water. He reasoned from this that all objects displaced water according to their *weight*. If he put the gold crown in water, it should displace the same amount of water as bulk pure gold. He was so overjoyed by the discovery that legend says he ran down the street, naked and still wet from his bath, shouting "Eureka!" which means "I've found it."

Incidentally, the king's crown did have silver alloyed with it.

Making a Specific Gravity Test. We can make a specific gravity test of a metal by first weighing a metal in the air with spring scales. Then the metal, still attached to the scales, is immersed in water. The metal, but not the scales, must be under the water and not touching the sides or bottom of the container.

Suppose the metal weighs 4 lb in the air and 2 lb in the water. Subtracting the two weights gives us a 2-lb loss. Now if we divide the 2-lb loss into the original weight of the metal—4 lb—we get 2.0, which is the specific gravity of the metal we are trying to identify.

A check of the tables listed under Chemical Elements in encyclopedias shows us that the specific gravity of aluminum with a copper alloy is 2.7 to 3.0. These figures rule out this alloy as our metal, but further checking discloses that aluminum-magnesium alloys have a specific gravity ranging from 2.0 to 2.5. This could be our metal.

When Specific Gravities are the Same

Sometimes you may find different metals with the same specific gravity. Pewter (an alloy of tin with lead) has the same specific gravity—7.0 to 7.6—as cast iron.

Pewter resembles silver; the color and weight would tell us that it is not iron and the texture would be different. If the two metals were such that color and texture would not identify them, then we could use a file test. The pewter would be softer than the iron.

SUMMARY

1. Metals differ greatly in composition and the way they react to heat.

2. Approximately 90 percent of a welder's work is with ferrous (iron content) metals.

3. Ferrous metals have a cube crystal form that undergoes a change with the metal's temperature is heated to 1325° F.

4. This temperature is the *critical point* at which the iron crystals (grains) rearrange themselves. The grains of low carbon steel begin to grow smaller. The grains of high carbon steel immediately grow smaller.

5. As the temperature rises, the grains of high carbon steel begin to enlarge. The grains of low carbon steel continue to diminish in size until they reach the upper critical point of 1550° F. At this point the low carbon steel grains begin to enlarge again up to the melting point.

6. Fine-grain ferrous metals are harder than large-grain ferrous metals. The natural grain of these metals can be altered by changing the cooling rate of the metal.

7. A molten metal, if allowed to cool slowly, will return to its original grain size. If suddenly chilled when

the grains are at their smallest size within the critical range, they will retain their small, hardened state until reheated above the critical point.

8. Rapid cooling, called quenching, is done by plunging the heated metal into water, oil, or similar cooling substances.

9. Quenching hardens metal. Annealing is an opposite process that softens metal to render it more easily worked. This is done by heating the metal and cooling it slowly in an annealing oven.

10. Most metals are alloys or mixtures of a base metal with varying degrees of other metals.

11. Metals must be identified to select the proper process, filler rod, and heat.

12. There are several tests that can be used to identify metals. They include the spark, ring, fracture, chip, oxyacetylene, and specific gravity tests.

NEW WORDS AND TERMS

Alloy To mix small amounts of a different metal with a base metal to improve its qualities.

Alloy Steel Carbon steel alloyed with metals like chromium, nickel, manganese, molybdenum, and others.

Annealing Softening metal and relieving stresses by heating it above the critical point and cooling it slowly.

Braze To weld with filler material that has a lower melting point than the base metal.

Carbon Steel Basic steel; iron with controlled amounts of carbon.

Chip Test Use of a cold chisel to remove slivers of metal to identify metal hardness types.

Critical Point The temperature 1325° F, at which point the atoms of iron are redistributed, causing grain crystals to diminish in size.

Critical Range The range from 1325° F to 1550° F, over which temperatures low carbon steel grains reduce to their smallest point.

Ferrous Metal Metal with an iron content.

Fracture Test The breaking of metal to inspect its interior grain pattern for identification of the metal.

Grain Crystalline pattern of solid metal.

Gray Cast Iron Pig iron alloyed with carbon, silicon manganese, sulfur, and phosphorous. A soft iron used for casting.

High Carbon Steel Steel with a high percentage of carbon.

Low Carbon Steel Steel with a carbon content of 0.05 to 0.30; also known as mild steel.

Malleable Iron White cast iron that has been softened by annealing for 40 to 60 hrs and then slowly cooled.

Medium Carbon Steel Steel midway in hardness between low and high carbon steels.

Molten Metal Fully melted metal.

Nonferrous Metal that does not contain iron.

Oxyacetylene Test Use of the oxyacetylene torch to melt a test puddle to check the weldability of a metal.

Pig Iron Smelted iron ore before any alloy metal is added.

Quenching The process of immersing hot metal in water or oil to cool it quickly to retain a specified grain size and increase hardness.

Ring Test A test for metal identification that depends upon listening to the sound pitch of a metal.

Spark Test A test for metal identification by observing the kind and color of sparks created by touching a metal to a grindstone. Can only be used with ferrous metals.

Specific Gravity Test A method of identifying a metal or alloy by measuring its specific gravity. Specific gravity is the weight of an object divided by the weight of an equal volume of water.

White Cast Iron Cast iron that is cooled quickly in the mold. Less carbon separates from the iron than with gray cast iron. It is extremely hard.

Wrought Iron Pure pig iron mixed with iron oxide and silica.

OXYACETYLENE WELDING

THE OXYACETYLENE PROCESS

Oxyacetylene, the fuel used in the oxyacetylene welding process, is a blend of two gases: oxygen and acetylene. Oxygen, an element that makes up about 20 percent of the earth's air, is necessary to support combustion—we cannot have fire without oxygen.

This can be proved by setting a can of gasoline afire. If a tight cover is placed over the burning container, thus shutting off all air, the fire will go out. Foam fire extinguishers put out a blaze by covering the fire and cutting off the supply of air.

ACTION OF OXYGEN AND ACETYLENE

Although oxygen is essential to support fire, it is not combustible itself. In many cases the normal oxygen in the air is sufficient to support a fire. In other cases extra oxygen must be supplied. The Boy Scout or the hunter who gets down on his knees and blows on live coals to make them flare up is supplying extra oxygen to make a weak fire stronger.

In other cases the normal (ambient) air may supply enough oxygen to make a substance burn, but extra oxygen will make the flame much more intense. Acetylene gas is such a substance. Ignited under ordinary conditions, acetylene gas will burn with a comparatively weak flame, but if mixed with pure oxygen, it will burn with temperatures up to 6300° F. This is hot enough to melt any known substance that the welder will be required to work on. Without extra oxygen, an acetylene flame would not provide a working temperature over 4000° F.

WHAT IS ACETYLENE?

Acetylene is an artificial gas. It is formed by the action of water upon calcium carbides. Calcium carbide—called just "carbide" by welders—is formed by combining coke and lime. A pellet of carbide, dropped in water, reacts exactly like an antacid tablet does when dropped in a cup of water, as shown so often in television commercials. Bubbles of gas are formed in the reaction.

In the acetylene process, this gas is drawn off to be used or bottled for future use. The bottles are steel cylinders equipped with a tank valve on top. They may be obtained fully charged with acetylene gas from a welding supply house.

Shops that use large quantities may have their own acetylene generators. These are known as carbide-to-water generators because the carbide is added to the water. Gas can also be generated by dripping water on carbide, but this is used only for generating small quantities of acetylene, such as for use in miner's lamps.

The carbide-to-water generator (Figure 4-1) works in this manner:

1. Carbide pellets, contained in a hopper, are dropped as required into water placed in a sealed container.
2. The pellets dissolve in the water, forming acetylene gas bubbles.

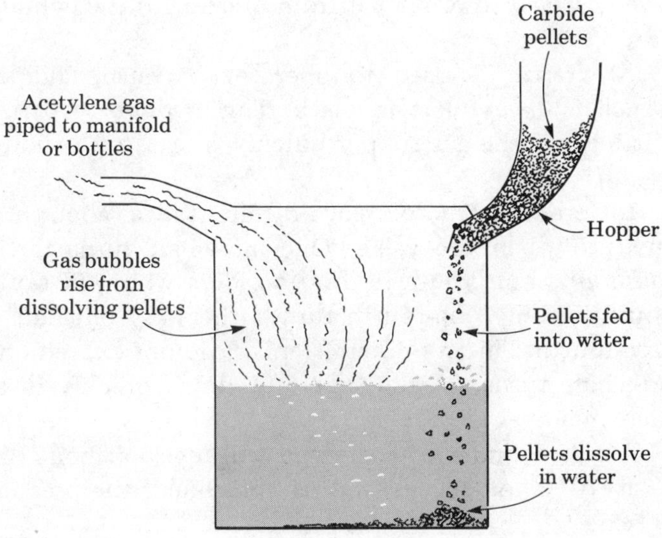

Figure 4-1. How a carbide-to-water generator works.

3. The bubbles are purified as they rise to the surface of the water. There they break and become acetylene gas.
4. The acetylene gas is drawn off into a manifold for use by a number of welders.
5. Acetylene gas is also converted from natural gas.

ACETYLENE CYLINDER

Acetylene in a free-standing form becomes unstable if subjected to pressure greater than 15 lb per square inch (psi) and can possibly self-ignite. Laws prohibit exceeding this pressure limit. However, these laws do not apply to *saturated* acetylene gas, which can be safely stored at pressures in excess of 15 lb.

Saturated acetylene gas is stored in a steel cylinder that has been filled with a porous matter saturated with an acetone liquid. When acetylene is pumped into such a tank, the acetone absorbs the acetylene gas and controls it. This allows a larger cubic volume of acetylene to be stored under pressure.

Controlling Cylinder Pressure

A valve at the top of the steel cylinder is used to close off the gas supply when welding is not in progress. It is also used to connect the gas regulator. Most products use a standard valve, but Linde Air Products recesses the valve in a well to protect it from breaking if the cylinder is dropped.

A wrench is used to operate the valve, and the wrench is *always* left in place. This is done to permit shutting off the gas supply quickly in case of a sudden fire.

Different instructors have different ideas about how much to open the valve. One-and-a-half turns is the *maximum*, but you will find welders who will say it should be only one-fourth turn. Others say one-half or three-fourths. In any circumstances, do not exceed one-and-a-half turns so you can shut down quickly in an emergency.

A full cylinder of acetylene will register about 225 psi at 70° F. As the volume is consumed, the pressure decreases.

Higher temperatures or fire in the shop will cause the pressure to rise. Safety fuse plugs made of low melting point metal are placed in the top and bottom of the tank cylinders to protect against pressure explosions that could rupture the cylinders.

The acetylene cylinder comes in three popular sizes for portable tanks. The large cylinder is rated at 300 cubic feet (cu ft) of gas. The next smaller size contains 100 cu ft, and the smallest is the 60-cu ft size.

In normal welding and cutting operations you will use about twice as much oxygen as acetylene. Therefore, a matched set of welding tanks would include approximately a 244-cu ft oxygen cylinder to a 100-cu ft acetylene tank (cylinder).

Figure 4-2. Pipes along the top of this picture carry acetylene and oxygen to the row of welding positions on the right-hand side of the picture.

Figure 4-3. This delivery truck has just been loaded with oxygen cylinders. The cylinders will be delivered to the welding shops serviced from this vendor.

Using a Manifold

A small welding operation will connect directly to the oxygen and acetylene cylinders. A shop that has a fairly large volume of work, but not enough to warrant using a generator, may connect several cylinders together to form a large cubic volume of gases to draw from.

If there is a centralized work area, this coupling is easily done by the use of a manifold. A manifold might be defined as a piping system that connects the cylinders together. Regulators are used at the end of the manifold to control the pressure from the cylinders.

A water-filled (hydrostatic) flame arrester is used in connection with the acetylene manifold. The flame arrester is a water trap that smothers any backfire which might lead back to the acetylene tanks. Backfire valves are installed at the regulator just before the hose is attached. The backfire valve is a one-way valve that closes if pressure is applied on it from the opposite direction.

PRECAUTIONS WITH OXYGEN AND ACETYLENE CYLINDERS

The transportation of cylinders comes under regulations of the Interstate Commerce Commission (ICC), which require that the oxygen cylinders be inspected and tested. This is done to ensure that the cylinders will hold a safe pressure without exploding. The first precaution in the use of oxygen cylinders is to check the ICC stamp to see if the cylinder has been properly inspected and tested.

The acetylene cylinders must always be used and stored in an upright condition. If they are not, the saturated acetone solution will drip through the valve and into the regulator and hose. The saturated acetone has a consistency similar to molasses and will ruin the regulator and the hose. It is a black substance, and if you ever notice it oozing from a tank valve or any other area, the cylinder should immediately be tagged unserviceable and removed from service.

Pressure Precautions

The pressure inside an acetylene cylinder is critical also. A full tank should register about 225 psi at 70°F and should be removed from service when the pressure falls to 50 psi. This is necessary because at lower pressure the saturated acetone may seep into the regulator and hose connections.

Acetylene cylinders should be kept out of the sunlight and away from stoves, radiators, or furnaces. If cylinders are stored inside, proper ventilation should be provided. They should also be kept dry and braced with a guard bar or chain to prevent them from being toppled over.

Lubricating Fittings

Petroleum and its products are very explosive in the presence of oxygen and acetylene. When it is necessary to lubricate any fittings on an oxyacetylene outfit, nonpetroleum base lubricants must be used. Glycerine, which may be purchased at a welding supply store or at a drugstore, is the most common nonpetroleum lubricant used by welders.

Checking for Leaks

Care must be taken that fittings do not leak. A heavy mixture of soap and water can be applied to the fitting with a small acid brush. If any gas is leaking under

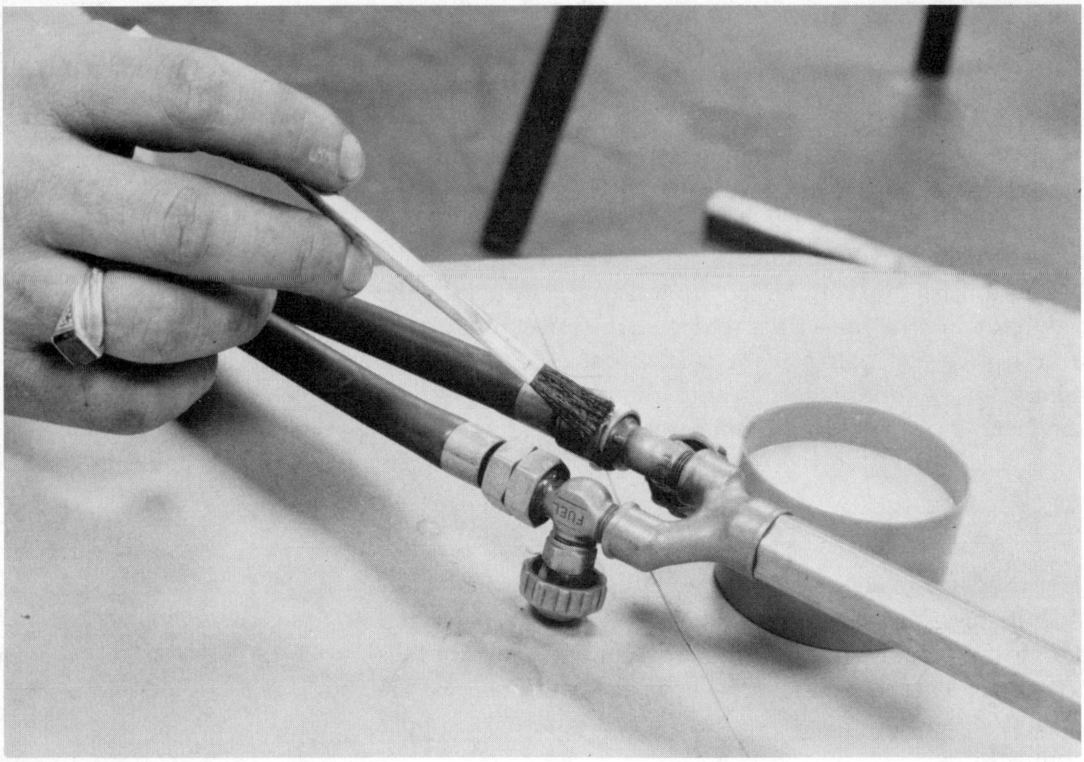

Figure 4-4. Fittings and connections are tested for leaks by brushing on a solution of liquid soap and water. Leaks will blow bubbles through the soap solution.

pressure, it will blow bubbles in the soap mixture. If such a leak is detected in a fitting, it should be carefully checked. The leak could be caused by the fitting not being tight enough. There could be improper contact with the inverted flare surface of the brass fittings. The trouble could also be caused by grit or dirt stuck in the flare contact surface.

OXYGEN

Except for low-temperature soldering of copper pipe, acetylene is always used in combination with oxygen. When oxygen and acetylene are burned under pressure in a welding torch they produce the hottest flame on earth. Arc welding can produce heat up to 10,000°F, but this is not a flame; it is an electric arc.

Normal (ambient) air is composed of about four-

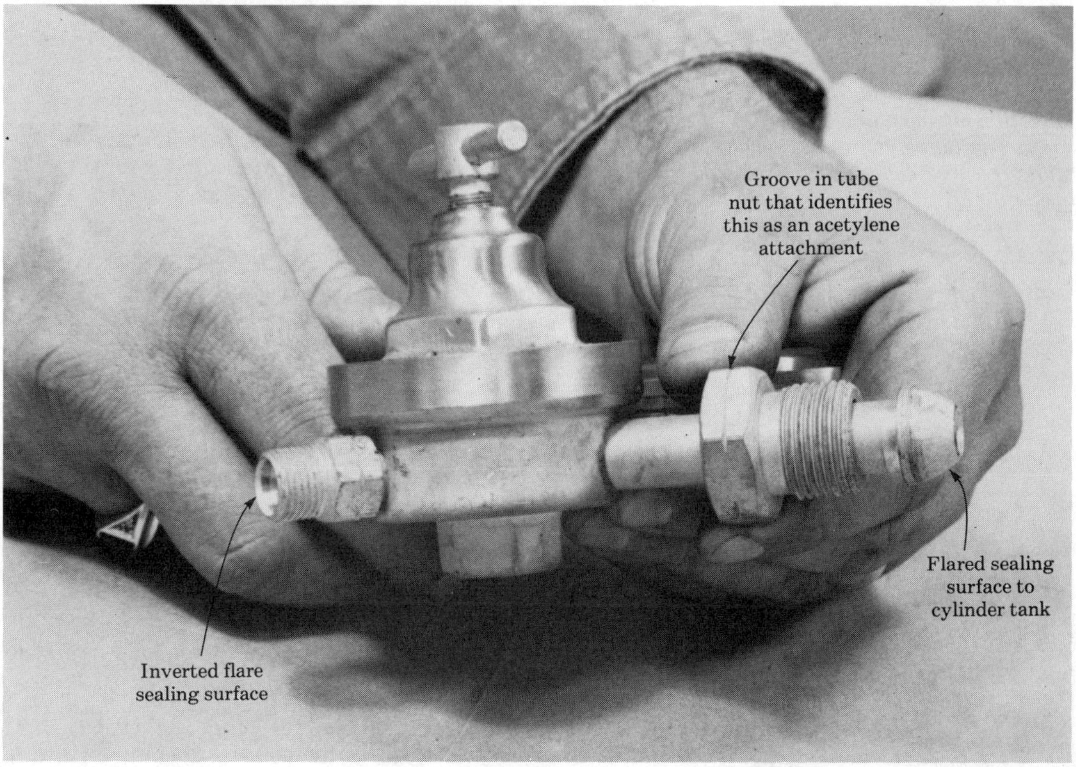

Figure 4-5. Connections for the regulators have an inverted flare seat. Connections should be checked for leaks after being connected to the cylinders. This is an acetylene regulator.

fifths nitrogen, and one-fifth oxygen with small percentages of argon, helium, carbon dioxide, and water vapor. The oxygen used for welding is concentrated from the air and is about 99.5 percent pure oxygen.

How Oxygen is Obtained

Oxygen is obtained by the liquid-air process. Every substance has a boiling point. Normal air is cooled below the boiling point of all the substances in it, then it is heated gradually until each substance reaches its individual boiling point. This method can be used to isolate the oxygen as it boils off. Oxygen boils (changes from a liquid to a gas) at −297.2° F. The oxygen gas thus obtained is 99.5 percent pure, which is satisfactory for both welding and medical use.

Figure 4-6. The vendor has large storage tanks that are used to fill the smaller cylinders. Pictured here are the storage tanks for nitrogen and oxygen gases. The tanks maintain their own pressure because of the gases inside them.

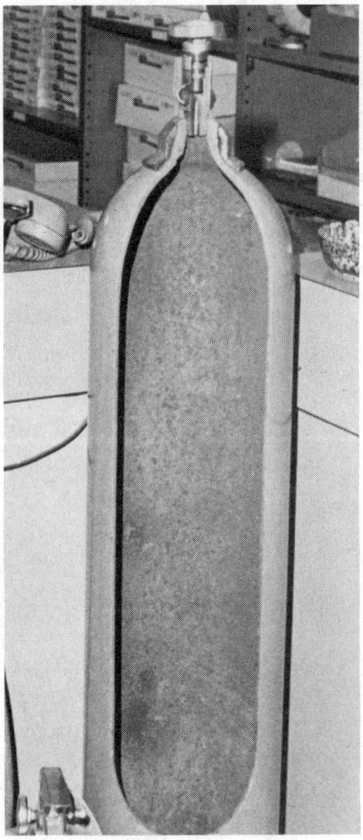

Figure 4-7. This oxygen cylinder has been cut away to show the valve and tank assembly. The cylinder walls are 3/8 in. thick to withstand the 2,200 lb of pressure.

Oxygen Storage Cylinder

The oxygen gas is stored for welding use in steel cylinders similar to, but larger than, the acetylene cylinder. The walls of the cylinder are 0.375 (3/8) in. thick. It is hollow inside, unlike the acetylene cylinder, and the gas in it is packed under pressure of 2,200 psi. Oxygen cylinders come in different sizes. Typical volumes are 244, 122, and 80 cu ft of displacement.

Acetylene and oxygen cylinder tanks are not often sold. They usually remain the property of the oxygen supplier, and the customer returns them for refill. There is no extra charge for the cylinders if they are returned within 30 days. Beyond this time there is a demurrage (late return) charge for each day beyond the limit.

Oxygen cylinders are subject to the ICC laws. Individual oxygen cylinders must be tested, inspected, and stamped every 5 years. There is good reason for controlling the use of these tanks: The contents are under heavy pressure; they have been known to rupture.

Ruptured oxygen tanks are like unguided missiles. They have been known to travel the length of a football field, ricochet off a wall, and still bounce far enough to kill a bystander who thought he was a safe distance away. The only exception to a 5-year inspection is when the tanks are set up as a 6-pack group of cylinders coupled to a manifold: then they are inspected every 10 years.

Figure 4-8. *This cylinder is being weighed to check that it is full.*

Figure 4-9. *A valve on top of the oxygen cylinder is opened all the way during use. It is closed hand tight.*

Cylinder Valve

The valve on an oxygen cylinder is opened all the way for use, instead of the one-half to one-and-a-half turns recommended for the acetylene valve. There is a wheel attached to the oxygen valve instead of the key wrench used on the acetylene cylinder.

The valve is protected by a cap that screws into place on the neck ring at the top of the cylinder. This protective cap should always be in place, except when the cylinder is properly secured for use. The cylinder should be against a wall or other firm object, protected by a chain to keep it from tipping over. If part of a portable unit, the oxygen and acetylene cylinders should be mounted on a wheeled cart, platform, or cradle and secured with a chain.

Safety Disk

Each oxygen cylinder is equipped with a small metal disk that will burst if pressure builds up too high in the

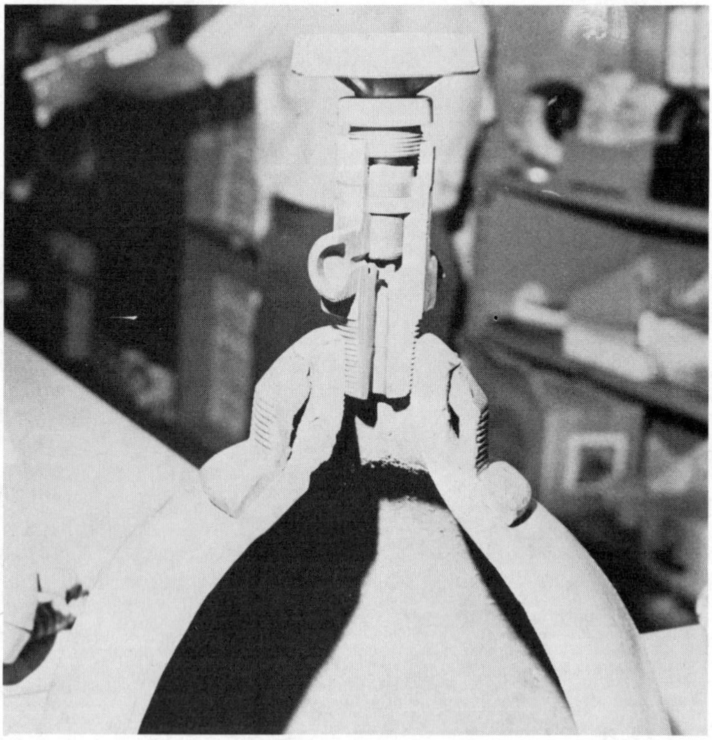

Figure 4-10. This close-up of the oxygen cylinder valve shows the double seat that seals the valve in the full open position as well as the full close position.

Figure 4-11. Both oxygen and acetylene cylinders are protected by steel caps that fit over the valve. This cap, called an "acorn" by welders, should always be in place when the cylinder is not being used.

cylinder. When this safety device bursts, the cylinder pressure is released through a nut that directs the pressure from the cylinder tank in all directions. This helps prevent the reaction of the released oxygen from turning the cylinder into a rocket.

Figure 4-12. Whether mounted against the wall or carried in a hand cart, oxygen and acetylene cylinders must be secured to prevent them from toppling over.

Figure 4-13. The safety valve permits slow dispersal of oxygen if pressure builds up too high inside the cylinder.

Bulk Use of Oxygen

The cylinder tanks may be coupled together by a manifold for increased volume capacity, as described earlier for acetylene. This is an easy way to provide larger cubic feet volume. A single regulator controls the manifold pressure leading into the shop; then separate regulators can be installed for each torch position. The operator for each torch then selects the pressure he needs for his particular job.

Very large quantity users may store liquid oxygen, which condenses the gas into a very small volume. The liquid is released into a manifold, where it becomes a gas. The small volume of liquid oxygen becomes a very large volume of gas.

Figure 4-14. These oxygen cylinders are all connected to a manifold that serves to fill the cylinders simultaneously. Each tank is filled to a pressure of 2,200 psi at 70° F.

Figure 4-15. A manifold is used to fill an oxygen cylinder with large cubic inch displacement.

Regulators

Each welding tank—both acetylene and oxygen—has a regulator to adjust the pressure of the gases from the tank to the torch tip. A regulator may look like two regulators because it has two dial faces. One of these dials indicates the cylinder tank pressure; the other shows the working pressure at the torch.

The pressure flow to the torch is controlled by turning the regulator screw clockwise to increase pressure or counterclockwise to decrease pressure to the torch tip.

Types of Regulators

There are two types of regulators. The first is the single stage, which is capable of high volume but subject to pressure variations. The second type, the double-stage regulator, maintains a more constant pressure setting but flows a lower volume than a single-stage regulator. If one uses the single stage, he must keep checking his settings to ensure that he is flowing the correct amount of oxygen and acetylene.

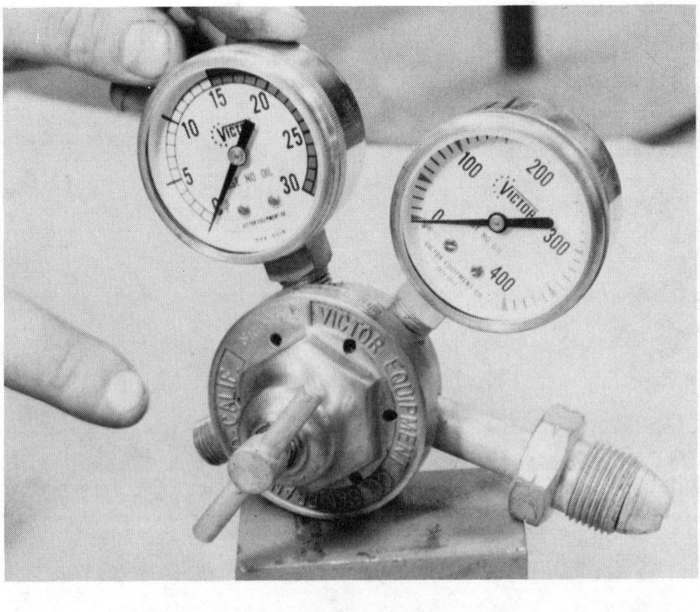

Figure 4-16. This acetylene regulator has two dials. The dial on the right indicates the pressure in the acetylene tank cylinder. The dial on the left shows the working pressure to the torch. The working pressure can be adjusted by changing the valve (indicated by the finger).

Figure 4-17. The oxygen regulator also has two dials. The left one indicates pressure (working pressure) to the torch. The right dial shows cylinder pressure.

SAFETY PRECAUTIONS

Working with fire and high pressures is dangerous, but welding is not a dangerous profession if you take proper precautions. A welding shop comes under the regulations of the National Board of Fire Underwriters, and your local fire department will inspect the premises periodically to ensure that the rules for storage of welding gases are being followed.

A very important, but often overlooked, safety rule is summed up in the Chinese proverb, "The beginning of wisdom is to call things by their right name." This is certainly true of welding. To make ourselves clearly understood by those we work with, we must learn the correct names for the tools and materials we use. Oxygen and acetylene are both gases, but we do not call either of them "gas"; neither do we call oxygen "air." Air to a welder means compressed air. Gas can mean anything from acetylene to propane. Oxygen is oxygen and acetylene is acetylene. They should be called such. Be specific and avoid uncertainty.

It is also well to remember that oxygen, although it does not burn itself, supports combustion. So, if your clothing becomes exposed to a full shot of 99.5 percent oxygen, do not get near a spark or flame because the oxygen will greatly increase the flammability of your clothing or anything else it may have saturated.

SOME RULES FOR SAFETY

Some of these safety rules have been mentioned before, but some are new. All should be obeyed with great care.

1. Do not try to "walk" a cylinder across the room. Use a truck or dolly.
2. Always check the ICC inspection and test record of each cylinder before using it. It is illegal to change these markings.
3. Protect oxygen cylinders from heat sources. Keep them dry. It also helps them work better if they are kept from ice and snow in winter.
4. Never use oxygen from a cylinder without a regulator.
5. Test cylinders for leaks by using the soapy water method, but do not try to repair the cylinders if a leak is found. Tag it and return the defective tanks to your dealer.
6. Do not stand in front of the regulator when first opening the cylinder valve. Turn the cylinder tank valves slowly at first and then gradually increase the pressure to the normal setting. This procedure prevents a sudden pressure on the diaphragm of the regulator, which could burst the glass lens of the regulator gauge.
7. Use no lubricants made with a petroleum base. Glycerine is a safe and satisfactory lubricant.
8. Periodically check fittings for leaks, using the soapy water test.
9. Let pressure off both gauges when the unit is not in use by first turning off the cylinder tank valve. Next open the needle valves on the torch one at a time, closing one before opening the other. This releases pressure in the regulator. It is advisable to remove tension from the regulator diaphragm by backing out (turning counterclockwise) the regulator screw until it is turning free. This relieves the regulator diaphragm from both the tank pressure and the spring tension exerted by the regulator screw.

Figure 4-18. Do not stand in front of the regulator when first opening a cylinder valve. Under certain conditions there is danger of pressure blowing out the gauge valve.

Thread Design of Fitting

As a special safety precaution, there is a common design on all welding equipment to prevent oxygen components from being interchanged with acetylene components. Each has its own thread design for attaching parts. The oxygen components have a right-hand thread design. These fittings are tightened to the right (clockwise). Acetylene components have a left-hand thread. They are tightened to the left (counterclockwise).

This means that you cannot fit an oxygen hose to an acetylene regulator or an acetylene hose to any part of the oxygen system. This precaution is necessary because of the different characteristics of the two gases.

There are also color codes for both oxygen and acetylene, which may vary with different manufacturers. One popular arrangement is to have the acetylene hoses red and the oxygen hoses green. The tanks with this system would have similar colored bands.

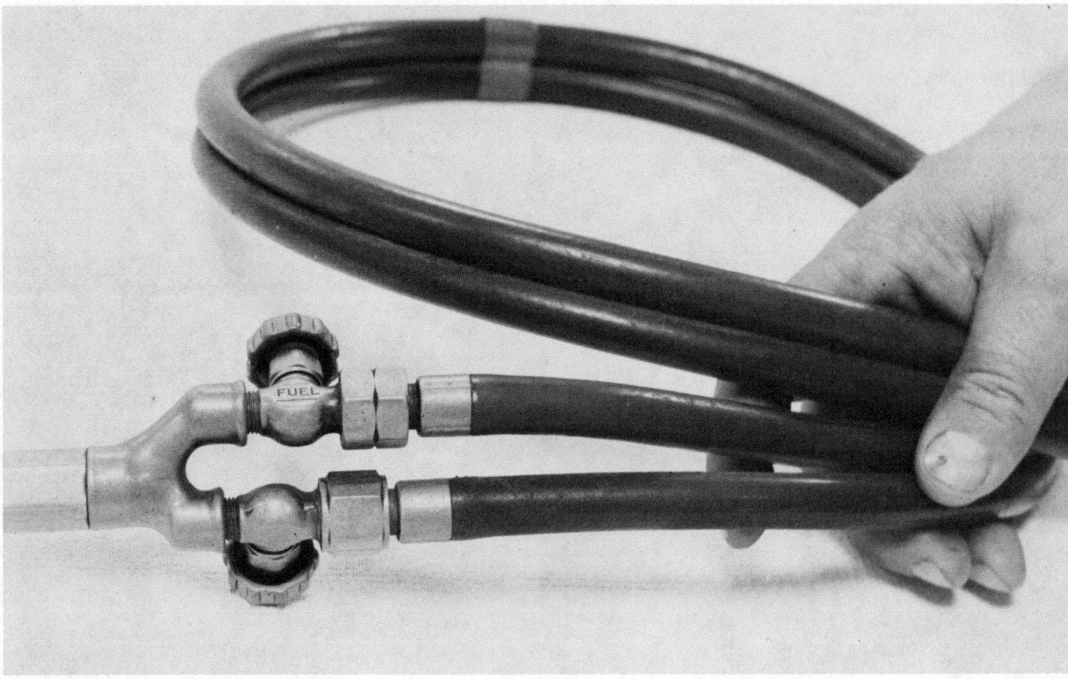

Figure 4-19. The oxygen hose (bottom) and the acetylene hose (top) have different thread designs on the fitting to prevent them from being interchanged. The fittings are also identified by the groove in the outside surface of the acetylene nut (top), whereas the oxygen nut (bottom) is plain.

SUMMARY

1. Oxyacetylene is a blend of two gases, oxygen and acetylene. Used in the proper proportions, the two produce a flame that may be as hot as 6300° F.

2. Oxygen is not itself combustible, but it is required to support combustion in other materials.

3. Acetylene is an artificial gas formed by the reaction of calcium carbide to water.

4. Large users of acetylene may use a carbide-to-water generator to manufacture their own acetylene, but smaller users purchase acetylene in steel cylinders from welding supply houses.

5. Oxygen is likewise supplied in steel containers similar to, but larger than, those containing acetylene.

6. Precautions must be observed in handling cylinders. They are always stored in an upright position and chained to prevent them from falling over.

7. The high pressure inside these cylinders makes them dangerous if the case is ruptured or a regulator is broken.

8. Each cylinder tank is equipped with a valve, and a regulator is attached for use. The regulator has two dials. One dial indicates the pressure in the tank. The second dial indicates the working pressure at the torch.

9. Petroleum products are very explosive in the presence of oxygen. Fittings must never be lubricated with petroleum products. Glycerine is the standard safe lubricant.

10. Leaks in fittings and hose connections are checked by brushing with a thick soap and water solution. Leaks will show through the soapy water as bubbles.

11. Extreme care must be taken in handling oxygen and acetylene. The safety precautions listed should be studied and restudied.

NEW WORDS AND TERMS

Acetone A chemical used in acetylene cylinder tanks to absorb free acetylene gas.

Acetylene A combustible fuel used with oxygen to create the hot flame used in oxyacetylene welding.

Boiling Point The temperature at which a substance changes from a liquid to a gaseous state.

Interstate Commerce Commission (ICC) A government agency that regulates the shipment of transportable items between states.

Glycerine A sirupy liquid used to lubricate welding fittings. It has a nonpetroleum base.

Manifold A piping system that connects several cylinders.

Oxygen A gaseous element present in the atmosphere. It is used in a 99.5 percent pure form in welding to increase the intensity of the acetylene flame. Oxygen is essential to the support of combustion. Fire will not burn without oxygen.

Pressure In welding, pressure is the force applied to acetylene and oxygen by its compression in the cylinder tanks. This pressure causes these two gases to flow through the regulator and to the torch.

Regulator A special valve system that controls the tank pressure and the line pressure to the welding torch.

Saturated Soaked to the limit of all that can be contained.

Volume The quantity of space within a container.

PREPARING METAL FOR OXYACETYLENE WELDING

Once metal to be welded has been properly identified, we have still another important step before we melt the material and form a weld bead. Proper preparation of metal for welding makes the difference between a strong weld and a weak weld.

CLEANING THE METAL

We must not forget that welding is a melting process. When we melt a dirty piece of metal, rust, dirt, and all the impurities are melted into the base metal to become part of the weld bead.

Recall from the discussion on alloying of metals that it takes only a small amount of a different metal or substance to alter the basic characteristics of many metals. A dirty weld is a weak weld.

There are several ways to clean metal for welding.

Cleaning Equipment

Sand blasting is an excellent way to clean rust and corrosion from metal, but such equipment is usually found only in larger welding shops. Although welders refer to this cleaning process as "sand" blasting, sand is but one of several substances that may be used. Abrasives, such as steel shot, seashells and walnut shells, and steel grit are used in these blasting machines.

Blast cleaning equipment is contained in a cabinet. Air pressure is used to blow the abrasives against the metal being cleaned.

Figure 5-1. This is a sand blaster. It cleans metal by directing particles of sand under great pressure against the dirty metal. The metal is placed inside the open compartment. The compartment is then closed, and the operator places his hands into the gloves. A window is used for viewing as the metal is cleaned.

Sometimes the corrosion can be chipped away with a chipping hammer. Hand grinders are sometimes used for cleaning, and so are wire brushes, sandpaper, and steel wool. Chemical cleaning is sometimes used on ferrous metals, but if chemicals are used, they must be completely washed off the metal before welding begins.

Proper safety precautions must be observed while cleaning the metal. Under no circumstances should you grind or chip metal without wearing a face shield or safety glasses. If chemicals are used, precautions must be taken to prevent burning the skin and eyes or breathing toxic vapors.

TYPES OF JOINTS

Any metal plate thicker than 1/4 in. should be beveled before welding to get good penetration. Otherwise the flame will not melt deep enough into the metal to give a solid weld. However, before going into the different ways to bevel metal, we must first consider the different types of joints.

Figure 5-2. Metal must be cleaned before it can be welded. The cleaning prevents impurities from getting into the weld bead.

Lap Joint

The lap joint is one of the most common joints. It is a very good one for a beginner because it requires no beveling or deep penetration. The joint is made by lapping one piece of flat metal over another, as the name implies.

The lap joint is welded along the overlapping edge as shown in Figure 5-3. For extra strength, the two pieces may be turned over and a similar weld made along the lap on the underneath side.

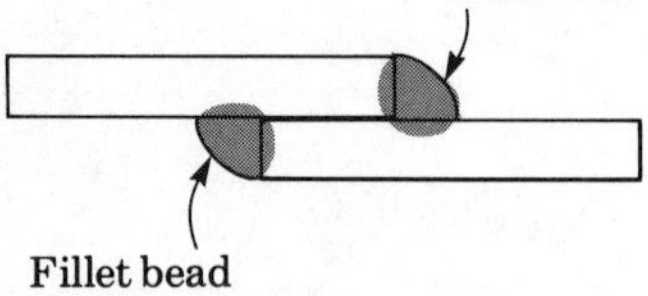

Figure 5-3. The lap weld is made by overlapping two pieces of metal and using a fillet weld bead at each butt.

Fillet Bead

The weld bead used for the lap joint is known as the *fillet weld*. A fillet bead is defined as one that joins two pieces of metal at a 90° angle. It could be called an inside weld; however, there is a difference between a joint and a bead. Basically, two beads are used with all the different joint designs—the fillet bead and the flat bead.

Figure 5-4. A fillet weld joins two pieces of metal at a 90° angle.

T Joint

The T joint is also an inside weld, requiring a fillet bead. The plates to be welded are placed at a 90° angle to each other in the form of a T. They are then joined with a fillet weld on one or both sides, depending upon the strength desired.

Thin plates may be butted directly together and welded. Thicker pieces of metal will require beveling in most cases. The beveling, as shown in Figure 5-6(*b*) and (*c*), is done on a power grinder or by flame cutting with the torch. Beveling will provide a stronger weld because the penetration of the fillet weld bead is complete.

A single 45° bevel is sufficient for metals up to 1/8 in. thick. Thicker metals may require beveling in a vee as shown in Figure 5-6(*c*).

Figure 5-5. The T joint joins metals at a 90° angle, requiring an inside fillet bead.

Root Pass

When metals are beveled, it is often necessary to make more than one pass to complete the bead to the desired

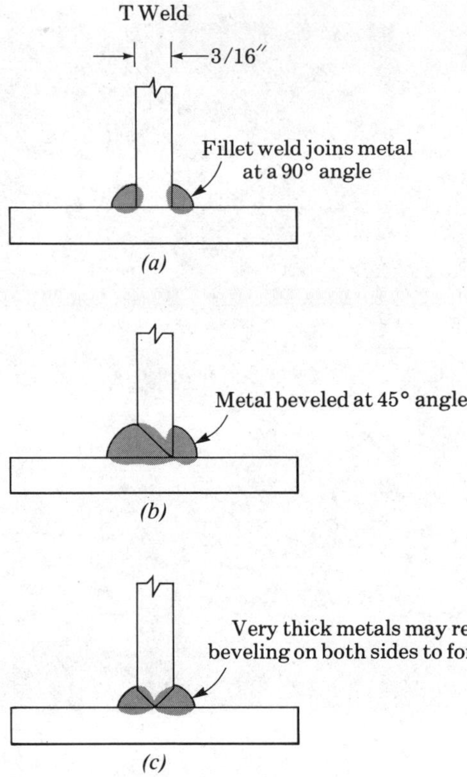

Figure 5-6. The T weld can be made by butting the metal (a), beveling at a 45° angle (b), or beveling with a vee bevel (c).

height and strength. The first pass is called the *root pass.* It is the most important because it serves as the foundation for the strength of the rest of the joint. Depending upon the thickness of the metal, it may be necessary to make only one pass. Thicker metals may require multiple passes.

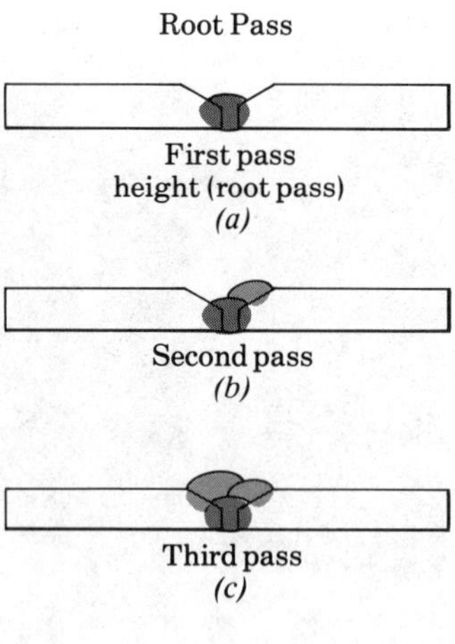

Figure 5-7. The root pass consists of joining the square butts with a strong weld (a) and then building up the remaining height with two successive passes (b), (c).

Butt Joint

The butt joint is for fusing together two plates butted end-to-end without overlaps or angles. The metal must be prepared by first cleaning away rust and corrosion.

There are three types of butt weld joints:

1. Closed
2. Open
3. Vee

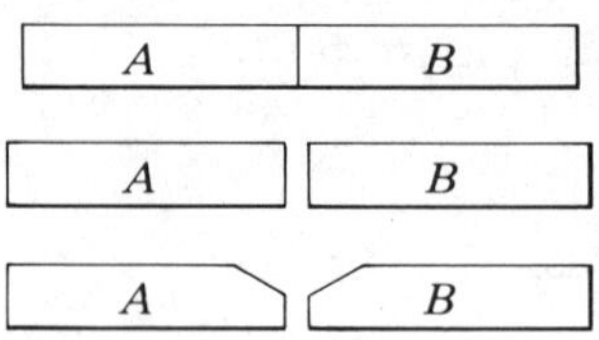

In the closed joint weld the two plates are in direct contact. Unless the plates are very thin, this may not be as strong a weld as the other two types of butt welds because the welding arc or torch flame will not be able to penetrate completely through the metal. Since there is no groove to fill, this type of weld would require little or no filler rod.

The open joint butt weld is made with the two plates separated by a distance up to the diameter of the filler rod, or slightly more, depending on the overall thickness of the metal. This separation is made to take care of expansion of the heated metal.

The open butt weld can be made by using a backup plate or a "keyhole" welding technique. The backup plate is a piece of metal that goes under the gap. It joins in the weld for complete fusion and is cut off later if required.

The keyhole technique is a practice in electrode manipulation. A small hole is allowed to form as the electrode penetrates the gap. The skill of the technique is in the control of the keyhole. The filler metal is added to the hole in a staircase technique, using a whipping motion to allow the molten puddle in the key hole to cool as the electrode is brought into the hole and then brought back out.

Excellent results can be obtained by using the keyhole technique. The main concern is the control of the puddle in the hole itself—if your whip technique is too long or too short, you will experience a molten blob for holding the electrode and heat too close, or the electrode will stick to the metal if it has been allowed to cool too much from pulling it out too long.

Beveling the Butt Joint. If the plates to be welded are more than 1/8 in. thick, the metal must be beveled.

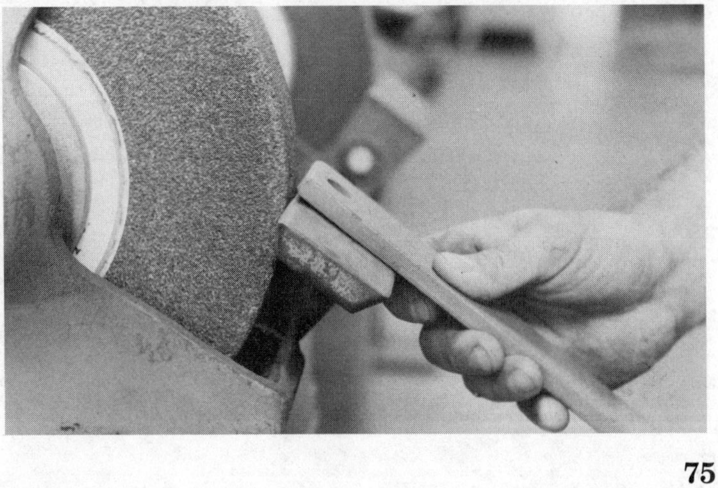

Figure 5-8. A power grinder can be used for beveling small pieces of metal that will be used for welding. The tool rest is set at the angle needed for the bevel.

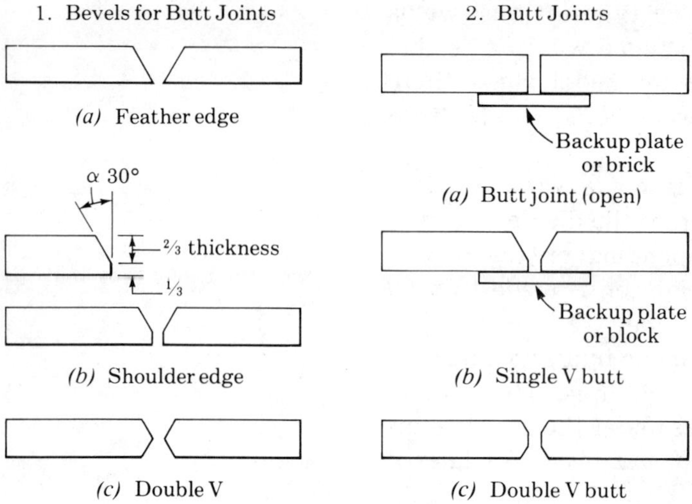

Figure 5-9.

This may be done in any one of three ways for the butt joint. One is the *feather edge*. Each side of the adjoining butts are beveled up to 30°. When the butts are put together, only the bottom of the vee thus formed will touch. Since the two pieces are each beveled up to 30°, the resulting vee would be a maximum of 60°. The groove is then filled with filler metal.

A second way to bevel is the *shoulder edge*. The two butts are also ground to a 30° angle, but the bottom one-third of the metal is left straight.

If the plates are over 3/8 in. thick, the two butts are beveled on both sides to make a *double-vee joint*.

The butt joint can be used to join curved sheets of

Figure 5-10. Butt welds can be made on shaped metal as well as flat metal. The seams of this large gas tank were made with butt welds.

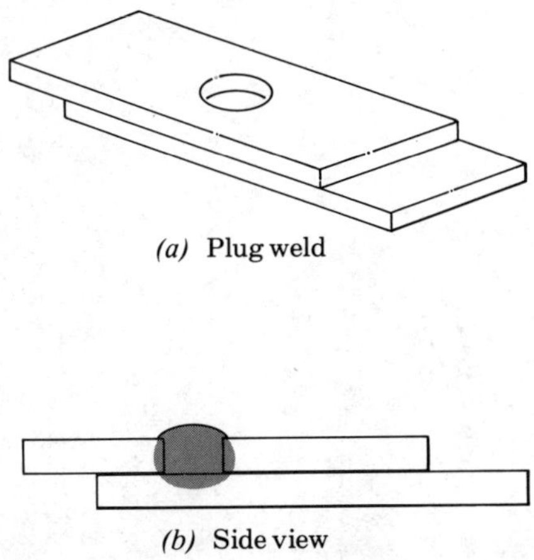

Figure 5-11. *Plug welds are an alternate to lap welds. (b) shows penetration of plug weld.*

metal as well as flat ones. Figure 5-10 shows butt joint seams in a large natural gas tank.

Plug Joints

The plug weld is an alternate to lap welds for fastening two pieces of metal together. In this type of weld joint the top lap of metal has a round hole in it (Figure 5-11). The weld is made through this hole. The surface of the underlying metal and the sides of the hole are melted with the torch. Filler rod is added to the puddle to fill the hole. The weld then literally becomes a fused plug fastening the two pieces together.

Preparation for the plug weld consists of cutting the plug hole or holes. These may be machine drilled or flame cut with a torch. The main precaution is to make the holes no larger than necessary. This will conserve filler rod and help prevent metal warp due to the use of excess heat.

Flange or Edge Weld

The flange or edge weld does not require filler rod. It fuses the upturned edges of two pieces of sheet metal, as shown in Figure 5-13. This weld is often used on thin sheet metal, where there is danger of the torch burning through the metal if a butt weld is used.

Preparation consists of crimping or bending the metal into the correct position.

Figure 5-12. The flange weld is made by bending the edges of thin metal at a 90° angle. The weld is made along the top of the seam. Filler rod is not necessary in this type of weld.

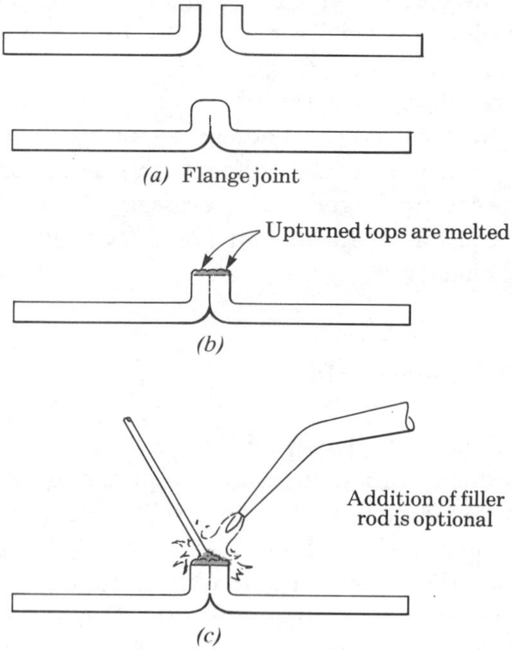

Figure 5-13. The flange or edge weld (a) is made by turning up the edges of thin sheet metal (b) and fusing the seam (c).

Figure 5-14. This inside corner weld has used two parallel beads.

Inside Corner Weld

The inside corner weld is exactly like the T weld except that the two plates are joined at right angles to their edges instead of one being in the center of the other. The fillet weld is used along the edge where the metal is to be joined.

As in the other welds, the edges are beveled if the metal is thick (Figure 5-14).

Outside Corner Weld

The outside corner weld can be done either with or without a filler rod. If the weld is to be made with a filler rod, the plates are prepared and set together in the manner shown in Figure 5-15(*a*). This is to bring together just the inside edges of the metal. The plates are then clamped in jigs to hold them in position, and the weld is made by filling the open area at the two butts with filler rod.

In the second method of outside corner welding, the two plates are prepared as shown in Figure 5-15(*b*). One plate is prepared to slightly overlap the other. This

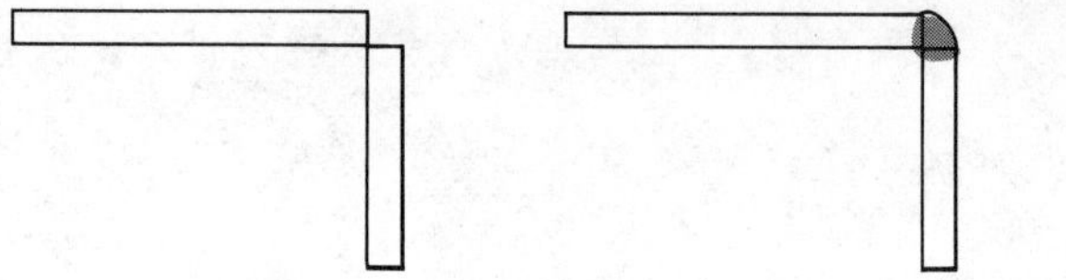

(a)
Fully opened
corner joint

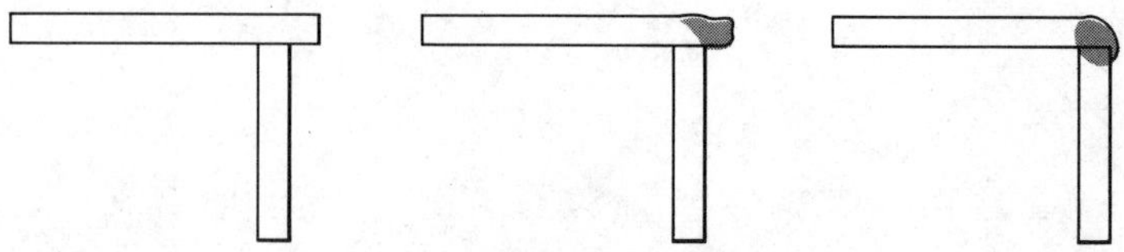

(b)
Overlap is used for filler
and no rod is necessary

Figure 5-15. The outside corner weld (a) and (b) can be made as shown with or without filler rod. (c) an outside corner weld made on an aluminum alloy with arc welding equipment.

overlap is then melted to form the weld bead as shown in Figure 5-15(*b*).

Still another way to achieve a strong outside corner weld is shown in Figure 5-15(*c*). In this technique one of the plates to be joined is beveled to ensure complete penetration of the weld.

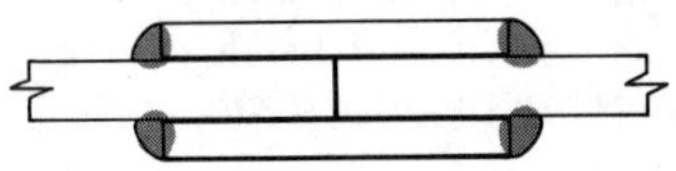

Figure 5-16.

Strap Weld

The strap weld is another form of the lap weld. Two plates are butted together and the joint covered with small laps of metal, as shown in Figure 5-16.

Round Stock Welds

The foregoing discussion dealt only with flat stock. Round stock is welded in the same manner and with the same type of joints. For example, welding a pipe to a flat sheet of metal is a butt weld joint, even though we weld around the pipe in a circle instead of in a horizontal seam. In a like manner, if we are welding two large solid rods together end to end, this is also a butt weld. If the rod diameter exceeds 1/8 in., we must bevel the butt ends of the rods in the same manner as we beveled the double-vee joints for flat plates.

The double-vee bevel for round stock ensures that our weld bead begins in the center of the rod.

Tack Welds

Tack welds are spot welds spaced at intervals to hold metal together. They are often used to keep metal plates securely in place while longer welding processes are going on, or they may be used where a complete bond is not necessary. If a full, continuous bead is not necessary, then it is a waste of time, welding rod, acetylene, and oxygen to lay a full bead.

JIGS AND CLAMPS

An important part of preparing metals for welding is to ensure that they are properly supported during the

Figure 5-17. A spot or tack weld is used to hold metal in place for longer welds or as a single weld when the finished article does not require great strength.

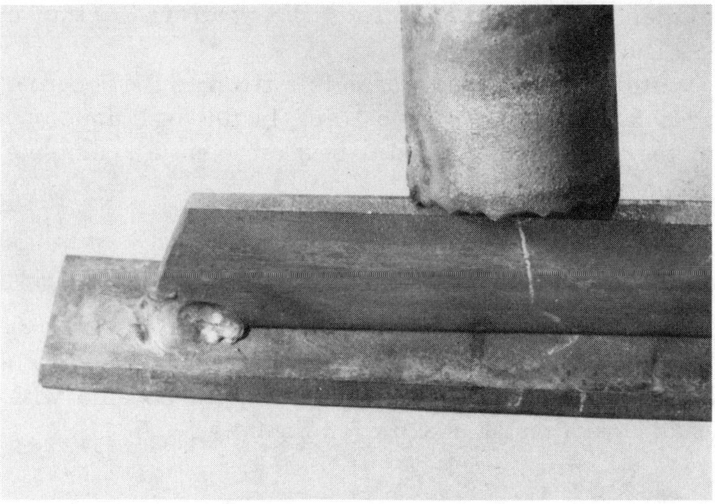

welding process. Heat causes metal to expand, and the subsequent cooling causes it to contract. This puts heavy strain upon the metal and is something that a welder must consider. This topic will be treated more fully in Chapter 9, *The Effect of Heat on Metal.* At this point we are concerned with getting metal ready for welding.

In some cases it may be necessary to place the base metal in a jig or under clamps (Figure 5-18) to prevent distortion. Sometimes we may need to provide a heat sink to absorb heat away from the base metal. The heat sink would be just a piece of metal placed under the

Figure 5-18. In some cases it is necessary to place the base metal in clamps or jigs to prevent distortion.

base metal that is to be welded, and it should be made of a metal which absorbs heat quickly. Copper is a good metal for a heat sink.

SUMMARY

1. Metal must be properly prepared before it is welded. Preparation includes cleaning, arranging the base metal plates in the proper position for the type of joint we intend to make, and beveling the butts, if this is necessary.

2. Beveling is usually indicated if the metal is more than 1/8 in. thick.

3. Weld joints include the butt weld, T joint, lap joint, plug welds, flange or edge welds, inside and outside corner joints, and the strap joint.

NEW WORDS AND TERMS

Backup Blocks Brick or metal placed behind a weld to prevent the torch burning through.

Bevel To cut the butts of plates at an angle. Beveled plates then form a vee trough at the joint, aiding full penetration of the welding flame.

Fillet Weld A weld bead that joins two pieces of metal set at a 90° angle.

Heat Sink A piece of metal placed behind the base metal to absorb heat and prevent buckling and distortion.

THE WELDING TORCH

The welding torch consists of a mixing chamber where the oxygen and acetylene are blended, two brass fittings for attaching the oxygen and acetylene hoses, and a selection of tips that can be attached to the mixing chamber. The brass fittings have valves for adjusting the line pressure of the oxygen and acetylene.

WELDING HOSES

Welding hoses, as mentioned earlier, are color coded. The oxygen hose may be green and the acetylene red, although this may differ with some manufacturers. There are different kinds of threads on the two valves to prevent an inexperienced welder from getting the hoses on the wrong valves. This is necessary because of the different composition of the two gases and their different pressure requirements. The acetylene hose carries less pressure than the oxygen hose, which must be stronger. Rubber is attacked by acetylene gas, so its hose must be of a material to withstand this chemical reaction.

Preparing New Hoses

New hoses have a small amount of powder inside to act as a drying agent to prevent moisture build-up during storage. This powder must be removed. It can be blasted out with gas, but compressed air is not recommended. Compressed air may contain atomized compressor oil. If this is left in the hose, it will explode or burn upon contact with oxygen.

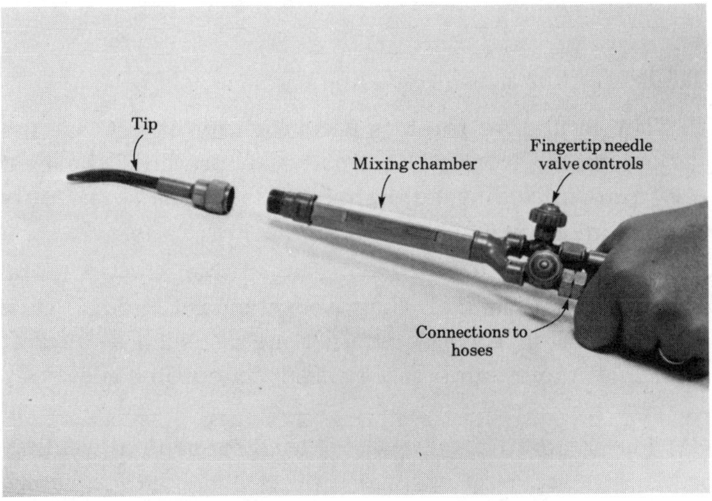

Figure 6-1. The welding torch has three basic parts: the fittings and valves, the mixing chamber, and the detachable tip.

Figure 6-2. These two backfire valves (a) are installed between the hoses and the fitting on the torch or cutting attachment. They prevent any flame during an explosion from traveling back into the hose.

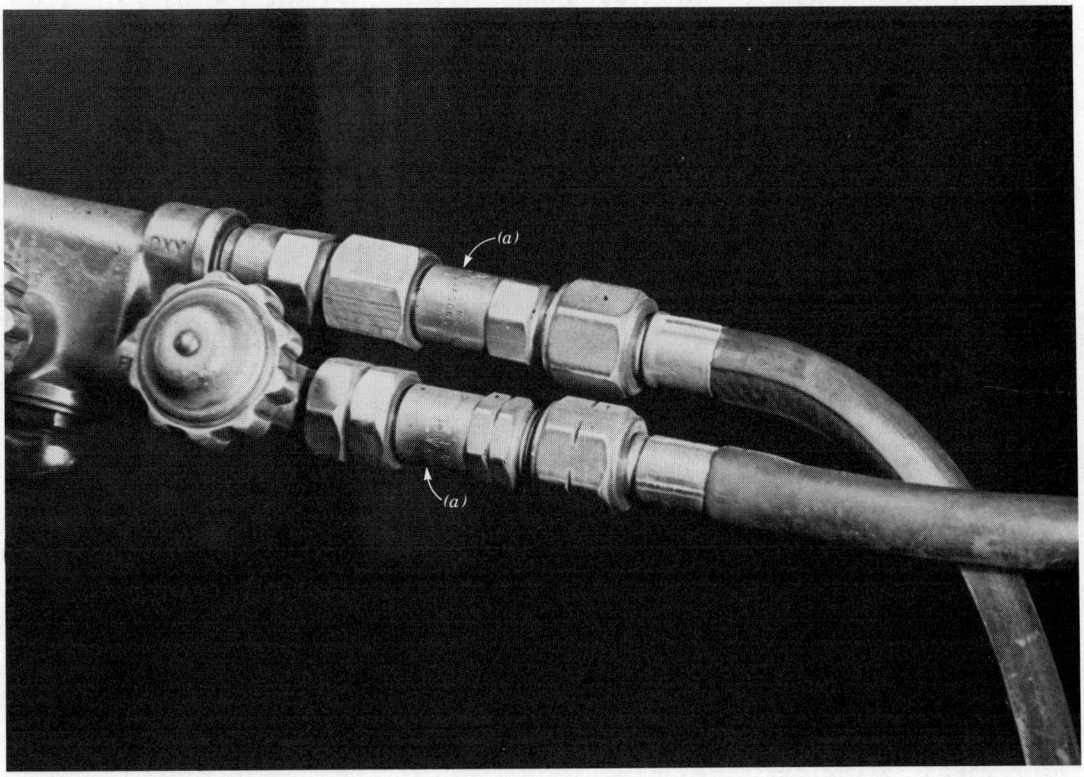

TORCH

There are four sizes of welding torches:

1. The miniature torch is used for tiny objects such as art metal work. The small size prevents the heat from unwelding adjoining joints that may be only 1/2 in. away.
2. The aircraft-type torch is larger than the art metal torch but smaller than the standard model. It is most often used on thinner metals, such as aircraft sheet metal and those used in automobile body work.
3. The standard-sized model is used for general welding.

   It is important to remember that all these models work on the same principle. They vary mainly in size.

4. The fourth type of welding torch is a slightly larger version of the standard model.

The capabilities of these torches overlap, so it is not necessary to buy all four sizes. A choice should be based on "What kind of welding jobs will I *mostly* do?"

Torch Tips

All the torches have a variety of interchangeable tips. The art metal torch has five tips and the standard model has seven. The size and diameter of the hole in the tip control the quantity of heat the torch can produce at the weld puddle. The larger the hole, the more cubic feet of gas the tip will flow. This will produce a flame with a larger heat output.

The difference in tip sizes permits you to control the heat for various thicknesses of metal, the type of metal (wrought iron, steel, aluminum, and so on), and the position of the welding joint. It will be necessary for a beginner to use each of the different size tips to learn by doing the effect each produces on various metals.

The seven tips used in common welding provide a wide useful range for the welder.

Tip Classification. The size classification of the tip is stamped on the brass collar of the tip; however, all tips

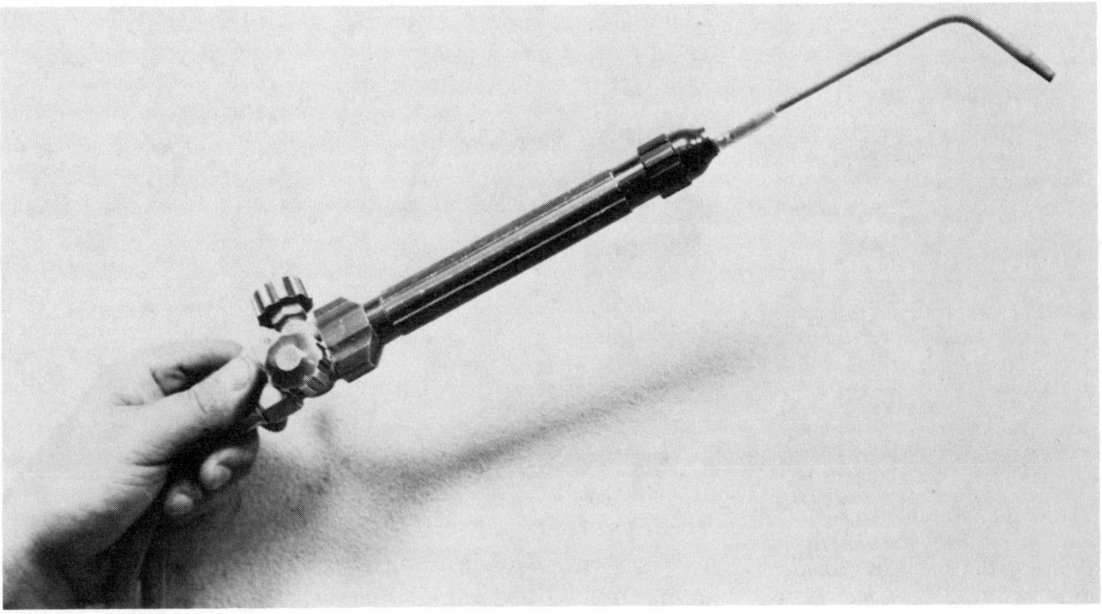

Figure 6-3. Welding torches come in different sizes. This one is used for small, delicate work.

Figure 6-4. The standard welding torch comes with interchangeable tips. Eight tips, ranging from number 000 to 5, are shown here. Other special-purpose tips are available.

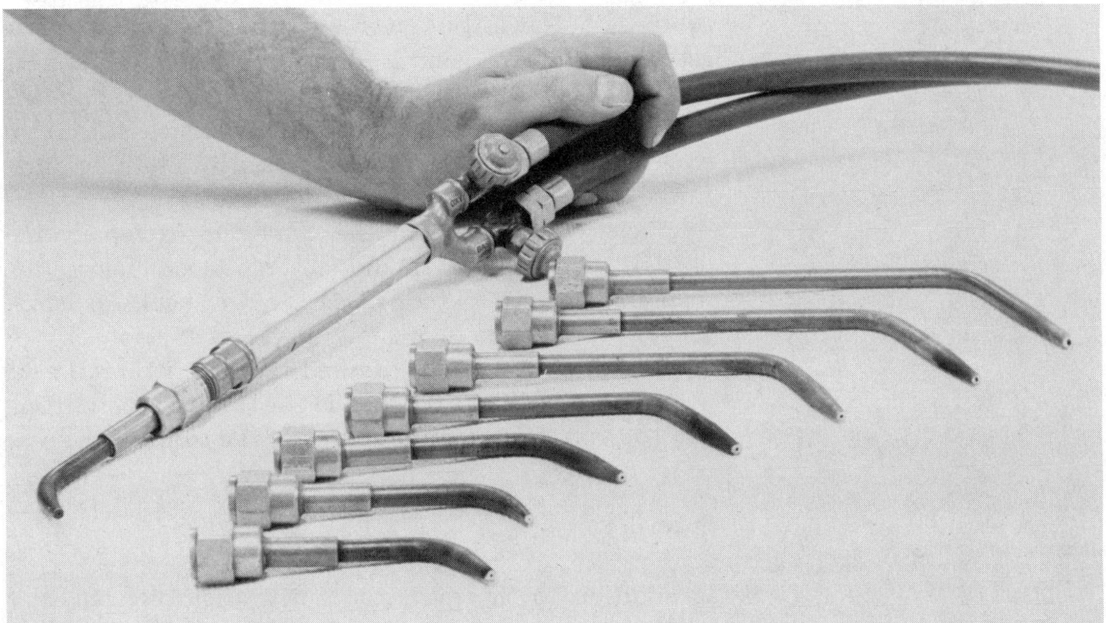

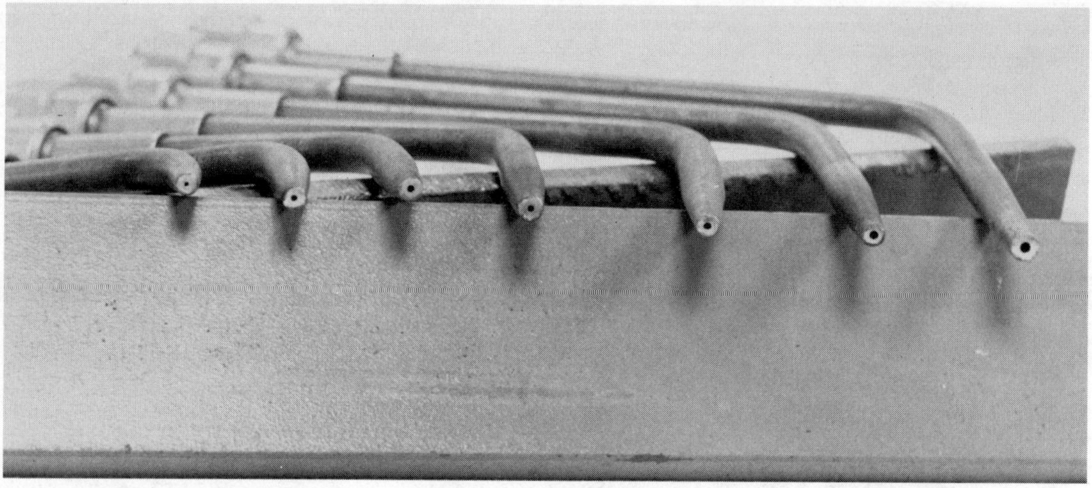

Figure 6-5. Welding tips come with differently sized orifices (holes in the ends) to provide different degrees of heat.

do not correspond to the same manufacturers' classification system. The manufacturers' tip classification will be compared to a decimal diameter. For example, Victor type 100 uses a 000 for their smallest classification. The next largest is 00 and then 0. The numbering changes at this point, and the next four tips are numbered 1 to 4, with 4 being the largest.

Tips can be quickly and easily exchanged. They are screwed off and on and hand tightened. A tight fit is ensured between the tip and the mixing chamber by an inverted flare connection with two neoprene O ring seals.

Dirt or grit in the fittings are the most common source of leaks, but worn O rings can also cause acetylene or oxygen to escape. Since the mixing chamber is used as the handle by the welder to hold the torch, any leak at this point could produce a fire that could burn the welder's hand or set his clothing afire. Loose connections can also produce leaks.

Tight connections in the torch are extremely important. The welder should never neglect making the soap bubble test, described earlier, to check for leaks. To repeat, since it is of such importance, a solution of liquid soap and water is brushed over the fittings. Leaks will blow small soap bubbles in the solution, permitting you to easily pinpoint the source of the trouble.

Special Tips

In addition to the regular tips with their different size holes, there are also some special design tips for

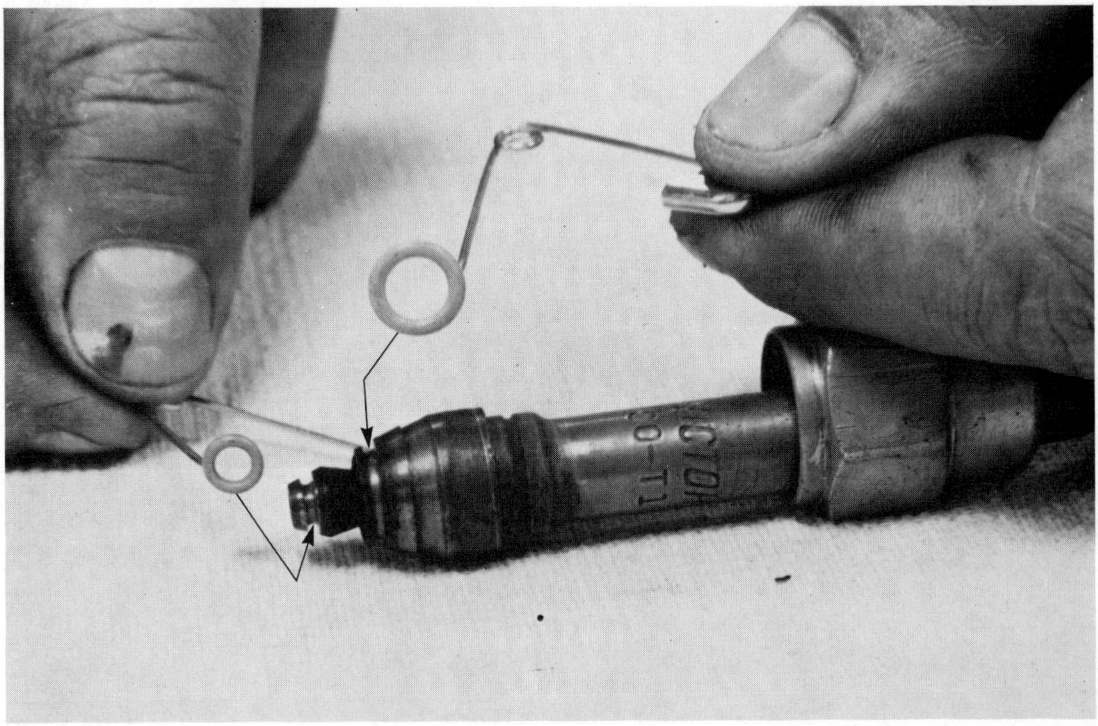

Figure 6-6. O rings on the fittings ensure gas-tight seals. As O rings become worn, they must be replaced.

specialized jobs. One such tip is the "rosebud." It has a large, multiflame burner end and is used for rapidly heating materials. This tip is very useful for bending and forming metals.

Flame Cutter

Another specialized attachment is the flame cutter. In addition to this attachment to the regular welding torch, there is also a special cutting torch. The cutting torch differs somewhat from the welding torch. In addition to the two needle valves to regulate the acetylene and oxygen flow from the cylinders, the cutting torch has a hand-operated lever to regulate the flow of additional oxygen.

In welding we melt the metal and let it reharden. Nothing is lost. In flame cutting we burn up the metal—or, as a scientific purist would say—we oxidize the metal.

How the Flame Cutter Works. The flame cutter tip has a central hole for the oxygen. This is surrounded by

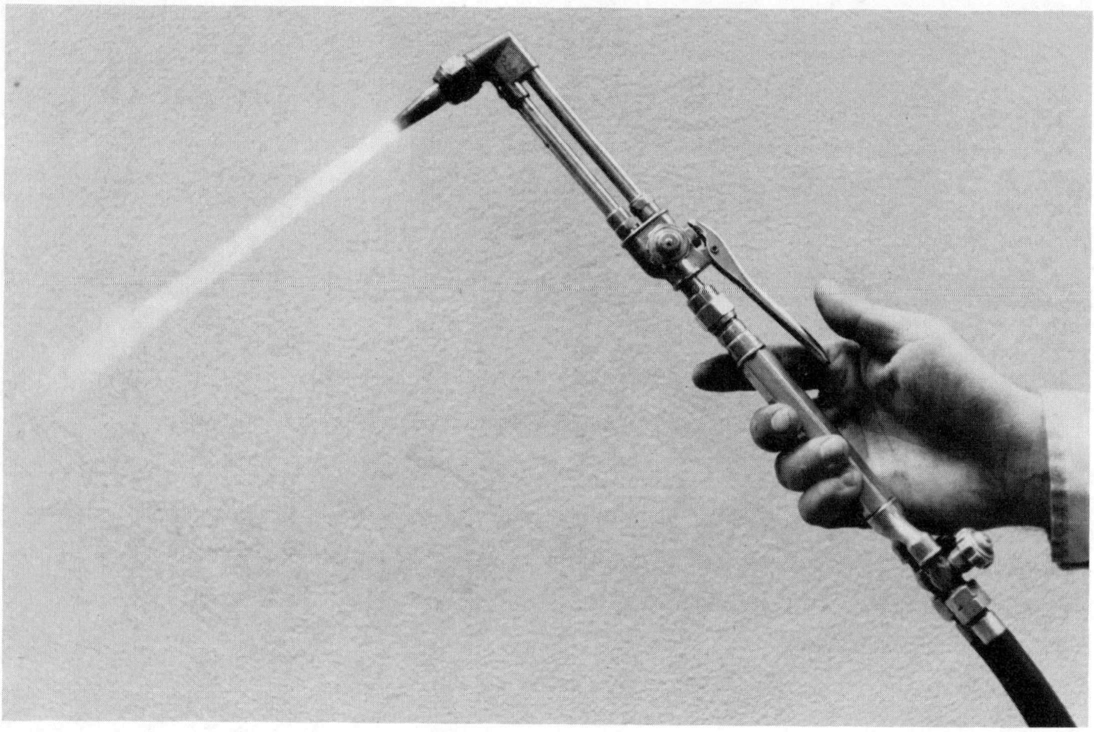

Figure 6-7. A cutting attachment permits the oxyacetylene torch to be used for quick cutting and beveling of metals.

a circle of smaller flame holes to preheat the metal. The preheat flame brings the metal to its "kindling" temperature (1200° to 1400°F), then a blast of pure oxygen from the center hole oxidizes the hot metal, carrying it away in a shower of sparks.

The flame cutting torch and the cutting attachment are valuable tools. They permit you to quickly cut holes in metal, bevel edges, and cut pieces to any desired shape.

Scarfing Tip

The scarfing tip is an attachment used with the flame cutting attachment to bring the flame down to cut close to a surface. The scarfing tip can be used for jobs like cutting the heads off rivets (Figure 6-8), although there is a flat tip that is better for removing bolts and rivets. The scarfing tip is most useful for making C grooves.

Other specialized tips are available. They are not used in ordinary welding but are very valuable—and sometimes essential—for unusual situations.

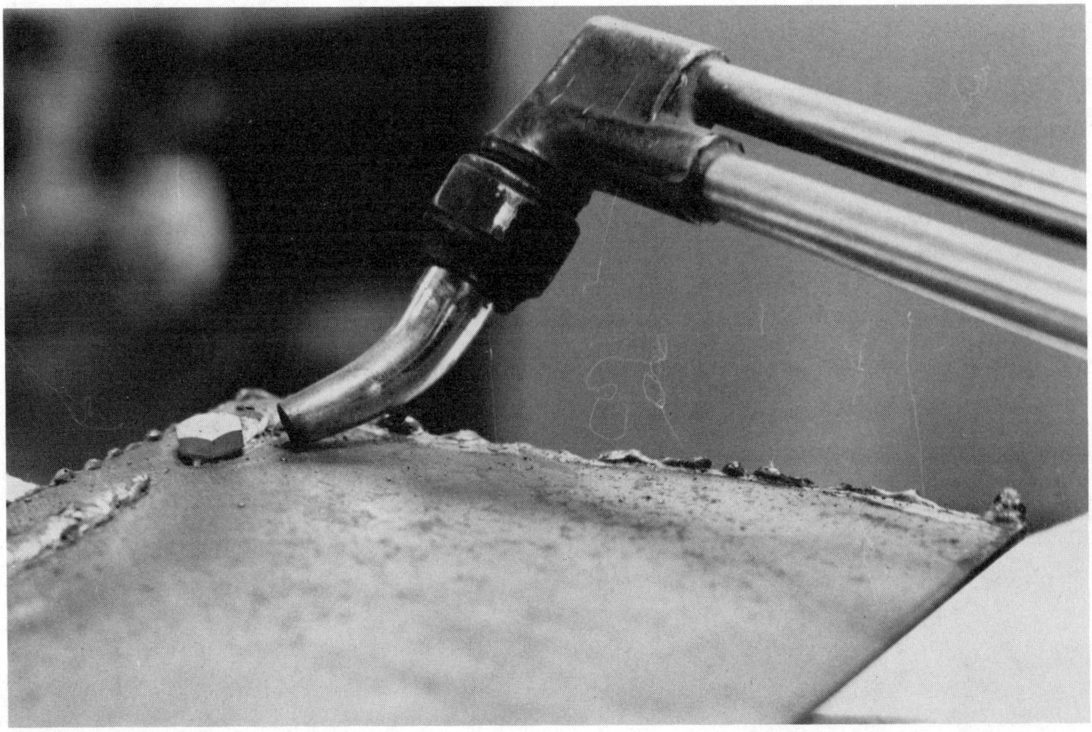

Figure 6-8. The scarfing tip permits direct fire along the surface of the metal to remove projecting material, like this screw head, without melting the base metal.

OXYACETYLENE SETUP

At this point we have identified our metal and prepared it for welding. The next step is to strike the flame at the torch tip. The following routine shows how this is done:

1. Check to see that the oxygen and acetylene cylinders are properly secured against the wall or in a portable cart. The safety chain must be in place.
2. Remove the protective cap from the neck ring and "crack" (slightly open) the valve on each cylinder. This will blow away dirt and grit that may have lodged in the valve openings.
3. Attach the regulators to the cylinders. Use a *nonpetroleum* lubricant if necessary to ease assembly. Tighten the regular fittings with a wrench.
4. Attach the torch hoses. You cannot get them wrong because one has a right-hand thread and the other has a left-hand thread. Tighten the fittings with a wrench (Figure 6-9).
5. Select and attach the proper size tip. *Tighten by hand.*

Figure 6-9. A wrench is used to tighten hoses to the regulator. All fittings are made wrench tight, except the torch tip, which is set hand tight only.

6. The size tip will depend upon the amount of heat needed. One manufacturer gives these recommendations:

Tip number	*Use with metal thickness*
00	1/64th
0	1/32d
1	1/16th
3	1/8th, and so on

Test for Leaks

The entire system must be checked for leaks after torch and hoses have been attached to the cylinders. Use the soap and water method, proceeding in this manner:

1. Slowly open the oxygen valve all the way. Stand to one side in case the sudden pressure blows out a defective regulator. Turn the regulator screw to read about 10 psi on the small figure scale. A full cylinder at 70°F will read about 2,200 psi on the large figure scale.

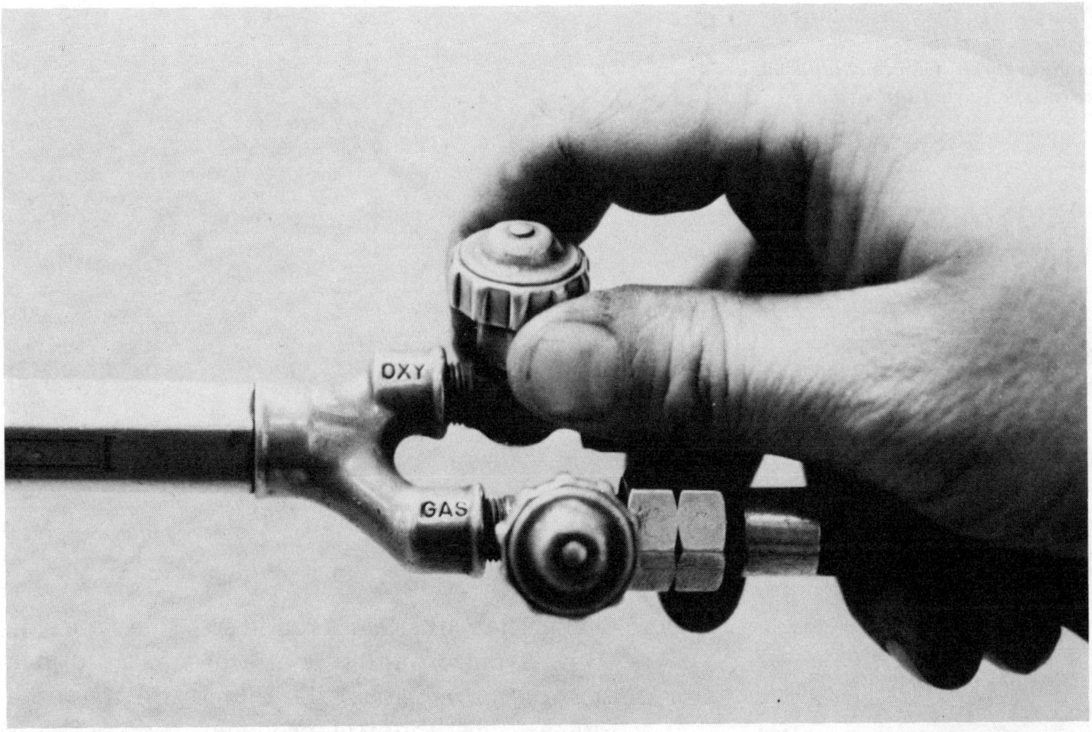

Figure 6-10. The needle valves are closed at the torch before making the leak test, to hold pressure in the hoses.

2. Tighten the needle valves on the mixing chamber. This contains the full 10 psi pressure between the cylinder and the mixing chamber of the torch. Now use the soapy water test for leaks in this pressure system.
3. Repeat the same procedure to pressurize the acetylene side of the welding system. However, *do not* open the tank valve all the way as you did on the oxygen tank. This is the burning gas, and we want to be able to cut it off in a hurry if an emergency arises. Open the cylinder valve only one-and-a-half turns. Adjust the regulator to read about 3 psi and close the mixing chamber needle valve. Now leak test the assembly the same way you did the oxygen side.

LIGHTING THE TORCH

After the system has been checked for leaks, you are ready to light the welding torch. Matches are too dangerous to use, so the lighting is done with a "striker"

Figure 6-11. The flint and steel striker is used to ignite the welding torch. Gloves should be worn when using the striker.

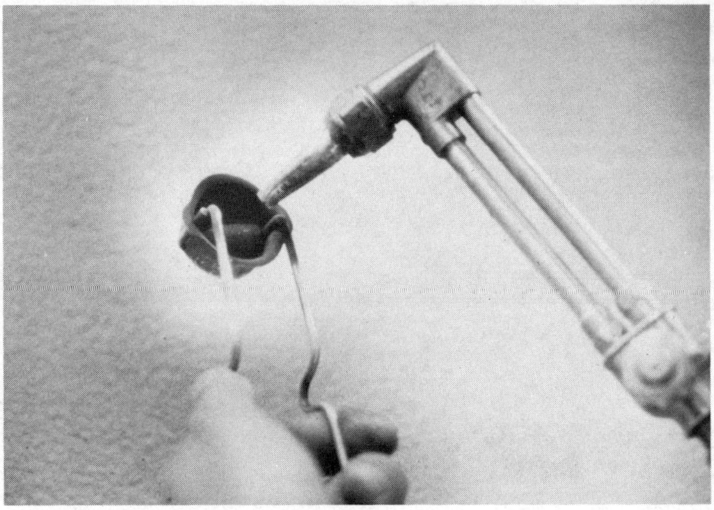

(Figure 6-11). The striker has a round steel file set inside a metal cup. A piece of flint is attached to a movable arm. A scissors movement of the thumb will cause the flint to scrape against the steel, producing sparks. This is the principle of the flint-and-steel pioneers used to start their fires. The striker keeps your hand far enough away from the tip so it will not be burned by back flare when the torch ignites.

To light the torch, open the acetylene valve slightly. Hold the striker at the end of the torch tip as shown in Figure 6-12. Press the striker arm to spark the flint against the steel. The acetylene gas, directed into the scraper cup, will ignite in a sudden flare of flame.

Acetylene Flame

At this point the acetylene, normally colorless, will burn with a sooty black flame. It will continue to do so until the gas flow is increased by opening the needle valve on the mixing chamber. Always direct the flame into the air, being watchful of people and things around you. It is very easy to burn someone or set fire to something.

Open the acetylene needle valve until the flame just begins to burn without leaving a black sooty trail. A small amount of black flame is all right. This adjustment produces the correct amount of acetylene gas flow for the size tip you have selected. If you change tips this

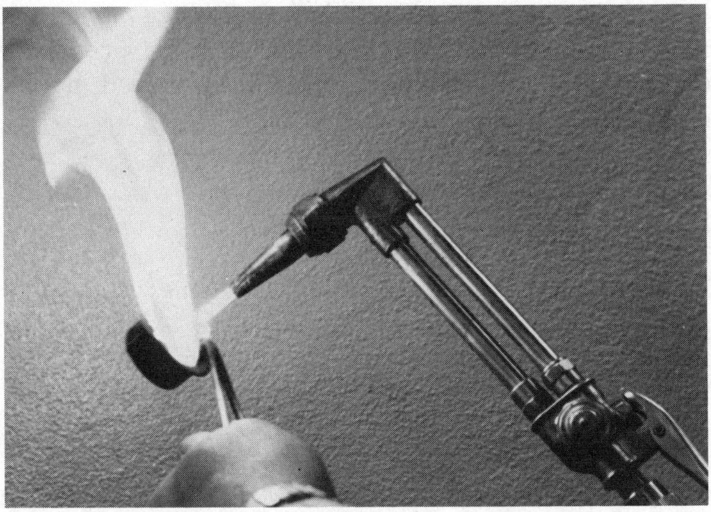

Figure 6-12. A flint and steel striker is used to ignite the oxyacetylene torch. Matches are unsafe because there is a flare of fire when the acetylene-oxygen mixture ignites.

procedure must be repeated and the gas flow readjusted at the needle valve.

Flame Types

The next step is to blend in the oxygen to produce the welding flame. Open the needle valve on the mixing chamber a small amount. You will see the long yellow acetylene flame begin to change to a shorter, bluer flame.

Continue adding oxygen until the flame looks like Figure 6-13. The flame is described as having three envelopes in the neutral flame position. The outer envelope is the larger, cooler flame. The second envelope is the smaller section of the larger flame. The third is the inner cone. This smallest envelope is the one that produces the hottest temperature. The tip of the neutral flame produces a temperature of about 5800° F.

If too much oxygen is added to the neutral flame, it will cause the inner cone to shrink. This produces an *oxidizing flame.* The oxidizing flame is used for special purposes, such as braze welding, but is not used in fusion welding. The excess oxygen will unite with the molten weld puddle, forming small bubbles that will weaken the weld. The oxidizing flame is the hottest produced by oxyacetylene. It reaches a temperature as high as 6300° F.

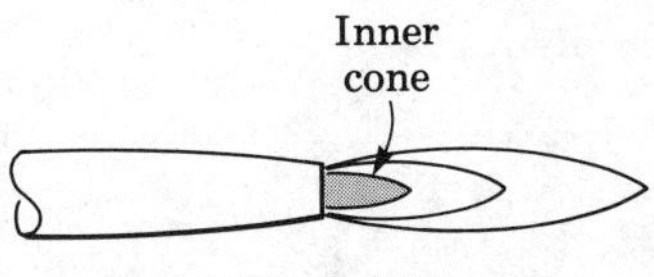

(a) Neutral flame

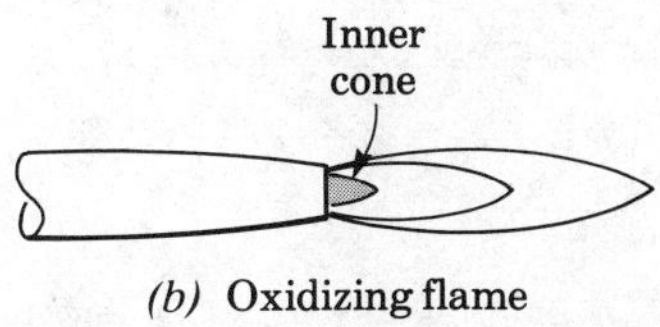

(b) Oxidizing flame

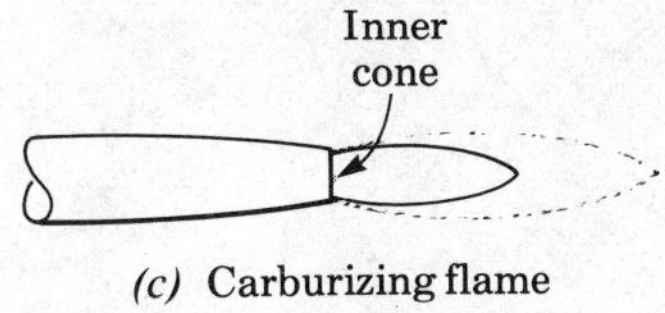

(c) Carburizing flame

Figure 6-13. The three flame types can be identified by the size of the inner cone. The flame has three envelopes. The third and smallest, the inner cone, is sharply defined in the neutral flame (a). It is sharply defined but smaller in the oxidizing flame (b) and is indistinct or invisible in the carburizing flame (c).

The third type of flame is the *carburizing flame*. It is the coolest of the oxyacetylene flames, producing temperatures around 5500°F. It is produced by an excess of acetylene in the fuel mixture.

Focusing the Cone

To clearly understand the carburizing flame, we must go back a moment to the neutral flame. Adjusting the flame can be compared to focusing a camera. When we first open the oxygen needle valve, the inner cone of flame contracts as we add more oxygen. When the flame focuses to a single clear image of the inner cone, we have the neutral flame. Adding excess oxygen gives us the oxidizing flame, but if we have excess acetylene instead of oxygen, we get a carburizing flame. In this condition the inner cone will not be a clear, sharply focused

Figure 6-14. The torch flame is adjusted until the inner cone (indicated by the circle) is sharply focused.

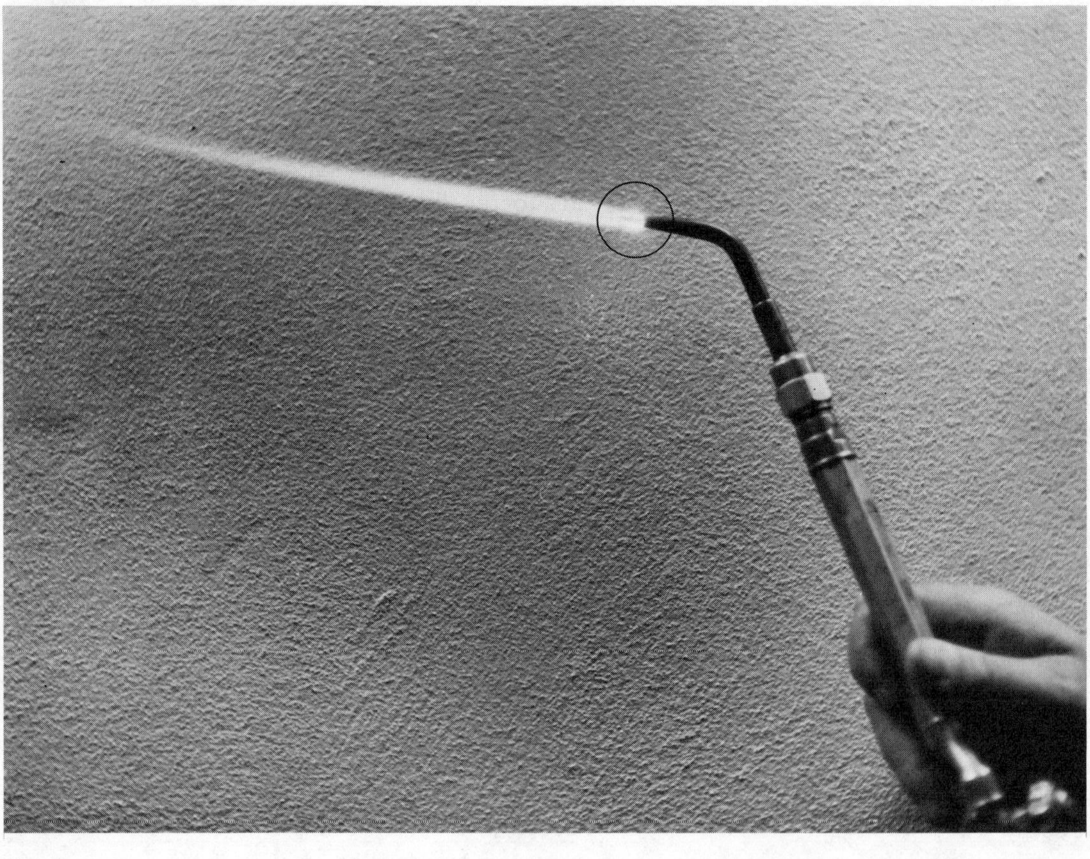

image. It is the flame stage before the neutral flame and is caused by not giving enough oxygen to the basic flame.

The carburizing flame is used for soldering and brazing with silver-base alloys where the lower temperature is desirable.

SHUTTING DOWN THE TORCH

It is as important to know how to shut down a torch properly as it is to know how to light one. In fact, there are times when it may be more important to shut down than to start.

There are two views on the subject. One holds for cutting off the acetylene first. The other closes the oxygen valve first. Because acetylene is the burning gas and the dirtiest, it is quicker and cleaner to close the acetylene valve first. This instantly extinguishes the flame. Then the oxygen valve is closed. Closing the two valves leaves pressure in the hoses between the needle valves and the regulators, so after the flame has been extinguished it is necessary to close the valves of the cylinders and to open the needle valves at the mixing chamber to bleed off the pressure in the hoses if no more welding is to be done at that session.

In the second method, the oxygen needle valve is closed first. This procedure produces a "pop" at the torch tip. Some authorities claim that this is hard on the equipment and even dangerous. Others claim that it blows out the built-up carbon from the tip.

The first method is preferred and should be followed by beginners.

CLEANING THE TIP

The orifice (hole) in the torch tip will become partially clogged. This may be due to carbon build-up, metal splatter, or both. The tips are easily cleaned by using a tip cleaner. The cleaner kit has a file to square off the tip surface and a set of gauged wires (wires of different sizes) with small abrasive teeth to clean tip openings.

To clean the tip, mount it securely. Then square the tip with the file. This removes accumulated splatter and carbon.

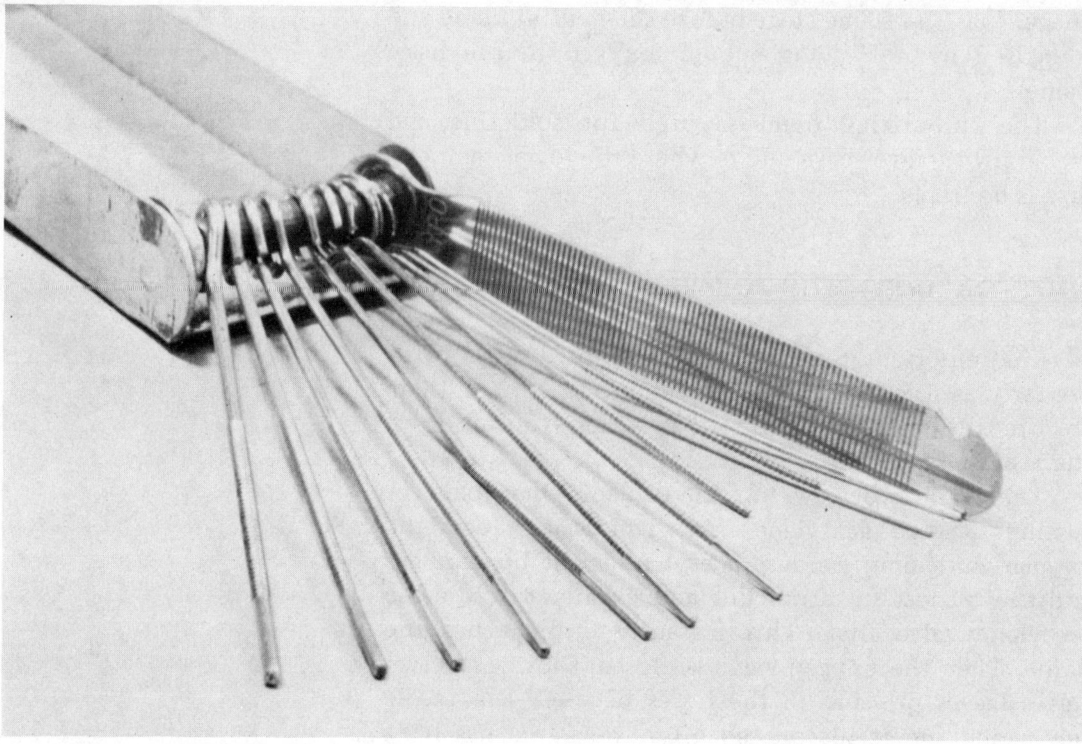

Figure 6-15. A tip-cleaning kit contains assorted size cleaning rods and a small file.

Next, select the gauge wire that matches the original diameter of the hole in the tip. The wire cleaner will fit snugly. Push the wire in and remove it. One pass in and out is sufficient. If you continue to ream the hole (push in and out), you will enlarge the orifice beyond its original size. This will change the heat value. It is very easy, for example, to enlarge a 000 size tip to a 00 size tip by excessive cleaning.

This procedure will clean the tip, allowing it to burn cleanly and evenly. A dirty tip will not focus into the single clear image necessary to obtain the required neutral flame. Sometimes a dirty tip will produce two ends like a snake's forked tongue. This condition prevents the control of the weld puddle and results in a weak, porous weld.

BACKFIRES

Inexperienced welders often have trouble with the torch backfiring or popping. Some causes of this are

1. Improper torch tip angle to workpiece

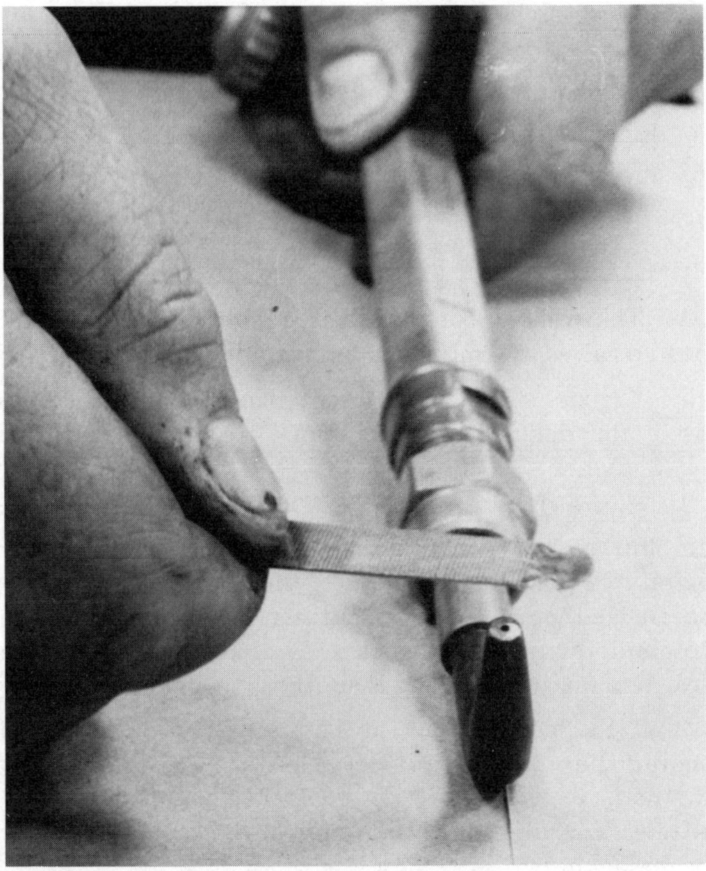

Figure 6-16. A small flat file is used to square the tip.

Figure 6-17. Correctly sized probes are used to clean the orifice (hole) in the welding tip. Care must be taken not to enlarge the orifice during such cleanings.

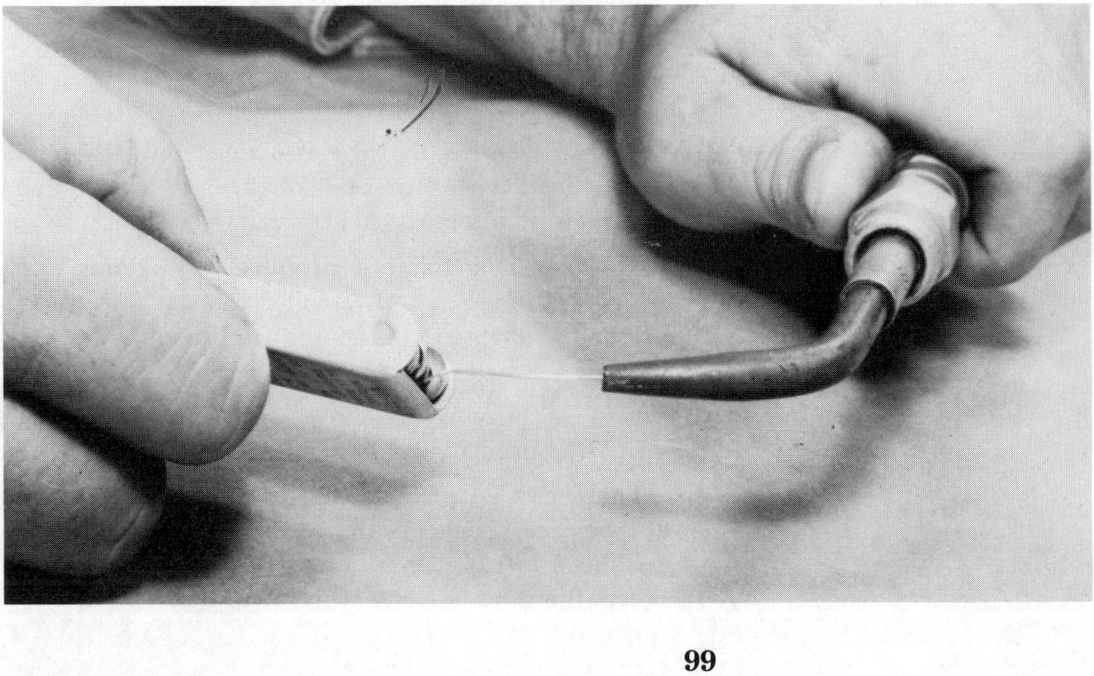

2. Loose torch tip or leaking O ring seal
3. Torch too close to the surface of the weld puddle
4. Use of a tip of the wrong size
5. Incorrect flame adjustment
6. Incorrect regulator pressure settings

SUMMARY

1. The welding torch has three parts: the fittings and needle valves, the mixing chamber, and the tip.

2. Torches come in different sizes, from small ones used for delicate, close-in work, to larger ones for regular work.

3. The size of the tip hole or orifice determines the amount of heat the tip can generate. In addition to a variety of graduated hole sizes, there are special tips for specialized purposes. The cutting attachment and the rosebud for heating larger areas are two examples. The scarf tip for close cutting is another.

4. Tight connections are essential, so connections should be checked for leaks with the soapy water method. O rings used for seals in the inverted flare fittings are subject to wear and can be replaced if they begin leaking.

5. A striker is always used to light the torch. Open the acetylene needle valve slightly, hold the striker in front of the escaping gas, and produce the spark.

6. The acetylene needle valve is opened until the flame burns without leaving a black, sooty trail.

7. The oxygen needle valve is opened until the inner cone of the flame envelope is focused sharply and clearly. This produces the neutral flame condition.

8. The oxidizing flame is produced by adding extra oxygen. The inner cone is smaller than in the neutral flame. The oxidizing flame is used for brazing.

9. The carburizing flame, a cooler temperature, is produced by having an excess of acetylene to the oxygen. It is used for soldering.

10. There are two ways to shut down a torch. One is to close the acetylene needle valve first, instantly extinguishing the flame, and then closing the oxygen valve. The second method reverses this procedure.

11. The torch tip must be kept clean. A cleaner kit consists of gauge wires for reaming carbon from the tip hole and a file for squaring the tip. Care must be taken not to enlarge the hole.

12. Backfires are caused by improper use of the torch, leaks, loose connections, and improper adjustments of flame, needle valves, or regulators.

NEW WORDS AND TERMS

Backfire A loud pop at the torch tip produced by an improper balance of oxygen and acetylene or from a poor tip connection.

Carburizing A flame produced by an excess acetylene-to-oxygen ratio.

Cutting Attachment An accessory for the torch mixing chamber that is used for flame cutting of metals.

Flashback A backfire that travels through the lines toward the cylinder tank. This can be dangerous unless stopped by a one-way check valve or flame arrester.

Neutral Flame Flame condition produced by adjustment of the mixing chamber needle valves to produce a single, clear, inner cone image.

Oxidizing Flame Obtained by allowing an excess of oxygen to the amount of acetylene. Used for brazing.

Rosebud A tip attachment used to heat metal.

Scarfing Tip A cutting attachment used to cut flush with a surface.

Striker A tool that produces a spark by rubbing flint and steel; used to ignite the welding torch.

Tip Cleaner A tool used to clean torch tips.

Torch Tip The attachment to the mixing chamber that regulates the flame temperature by providing different size holes which permits more or less acetylene to flow.

WELDING RODS

We previously stated that metal joints often have to be beveled before welding in order for the welding flame to penetrate deeply enough into the joint for a strong weld.

ROLE OF FILLER RODS

If a single-vee butt joint were to be welded simply by melting the sides of the joint and permitting the molten steel to flow together, we would get the effect shown in Figure 7-1(*b*). There would be a deep depression where the base metal was vee'd. The actual weld area holding the two pieces together would be slight. The joint would be weak and would fail under a heavy load.

We overcome this difficulty by adding additional metal to the puddle during the welding process. This filler metal is furnished in the form of welding rods. The welder melts these rods into the weld joint by dipping one end into the molten puddle as he welds. This provides the extra metal necessary to fill the depression which results [Figure 7-1(*b*)] when filler rods are not used. The final weld then looks like the one diagrammed in Figure 7-1(*c*). The weld has penetrated all the way to the bottom and has filled the entire groove between the two pieces of metal. The two pieces have now been fused into a single piece as strong or stronger than the original metal.

TYPES OF RODS

(a)

(b)

(c)

Figure 7-1.

Filler metal must be as close as possible in composition to the base metal being welded. The principle of welding is to create a weld that is as near as possible to the original metal. We can do this only if the metal we add to the weld is exactly like the base metal.

For example, if we have a section of mild steel to weld and we use a welding rod of high carbon content steel, we would have a joint that would be harder and more brittle than the rest of our metal. This is due to the greater carbon content of the metal.

To ensure that welding rod manufacturing is standardized, rods are made under specifications controlled by the American Welding Society. The Society requires that rods are made in sufficient varieties to correspond to the composition of common base metals.

Most filler rods used for oxyacetylene welding are uncoated. Coated rods, used in arc welding, give off a gas that surrounds the arc crater. This helps prevent oxidation of the molten metal. Oxidation (the combining of oxygen with the metal) can be controlled in oxyacetylene welding by an adjustment of the flame. This makes coated rods unnecessary.

SELECTING A ROD

Filler rods are selected by diameter and carbon content. The carbon content is to ensure that the rod is as close as possible in composition to the base metal.

The diameter of the rod, which ranges from 1/32 to 1/4 in., is chosen to most closely match the thickness of the metal to be welded. If the diameter of the rod is too small, it will require keeping the puddle molten longer to feed in enough metal. The extra heat could melt through the base metal or change the grain structure of the metal.

There is also a copper-coated low carbon steel rod. The copper has an advantage when used with dirtier metals because it increases the ease of the weld puddle manipulation.

Filler rods are generally purchased in 30-in. lengths, although there are some exceptions. The rods are blunt at each end, but they can be dangerous because of their small diameter. It is a good idea to bend the top end in a

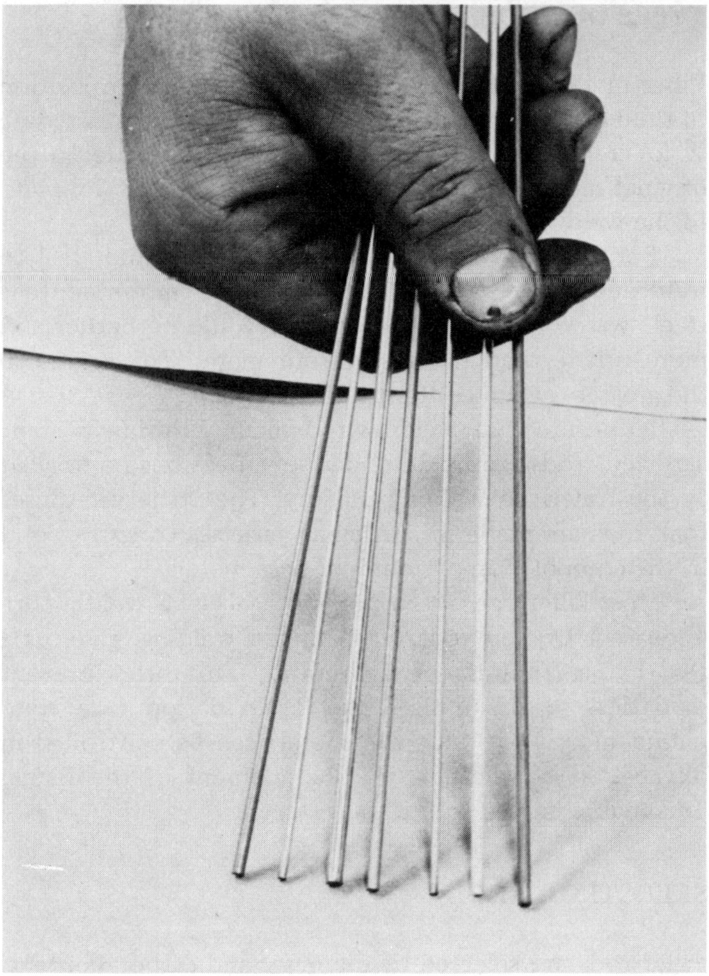

Figure 7-2. Filler rods are graded according to diameter size and carbon content.

loop. The rounded end will then be less of a danger to others who may be working closeby.

ROD STORAGE

Filler rods should be stored in a cool, dry place. This will prevent them from rusting. Rust, which is an oxidation of iron (combination of oxygen and iron) will contaminate a weld bead, resulting in a weak joint.

In a large shop filler rods are usually placed on a storage shelf. The shelf is divided to hold rods graded according to group carbon content. Within each group, the boxes are arranged according to the diameter of the rod they stock. This arrangement permits easy access to the right rod without loss of time.

SPECIAL FILLER RODS

We have been discussing ordinary low carbon rods used in oxyacetylene welding. In addition to these, there are special rods. One of these employs higher carbon content metal and is used for rebuilding surfaces or hard facing. Rebuilding might include restoring an entire broken gear tooth or restoring worn parts to their original size. Hard facing usually means to cover a metal surface with a harder coating of metal to make it withstand wear. Gears, plowshare points and edges, propellers, sliding joints, and other metal parts that encounter hard wear are helped by hard coating.

These parts may be of several different base metals that have been designed for specific wear properties. The welder must consider each repaired part differently. The welding rod chosen must give the best possible surface for that job and be compatible with the base metal.

BRAZING AND SOLDERING RODS

Iron and steel rods, with their varying carbon content, are for use with ferrous metals. Aluminum requires an aluminum rod and aluminum flux designed for this metal. Flux is a material used with the welding process to dissolve or help remove oxides and other undesirable substances. Flux combines with the impurities and either causes them to boil off or rise to the top of the weld bead as slag that can be chipped away when it hardens after cooling.

Silver soldering commonly uses two filler rods with the oxyacetylene process. These rods, along with brazing rods, will be more fully treated in Chapters 11 and 12.

Brazing, which is done with the oxyacetylene torch with an oxidizing flame, uses a nonferrous (no iron) filler metal that has a melting point lower than the base metal being welded. The broken pieces or joints are closely fitted together. The melted rod is drawn between the parts by capillary attraction. Capillary attraction is a surface tension that causes a liquid to be attracted to a solid. The base metal is heated above 800° F, which causes the rod to melt, but is well below the melting point of the base metal.

Braze welding also uses nonferrous filler rods. It

differs from straight brazing in the way the nonferrous filler is distributed. Brazing distributes the filler through a tight joint by capillary attraction. Braze welding, according to the official definition of the American Welding Society, deposits nonferrous filler in a groove, slot, fillet, or plug weld. The filler is melted, but the base metal is not.

HOW TO HOLD THE FILLER ROD

It is necessary to hold the filler rod in the proper manner to get a good weld. In the beginning this will be awkward, but soon your left hand will begin to follow what your right hand (holding the welding torch) does. Coordination between the two hands is essential.

It is usually best to begin with a thin metal. Some recommend 12 to 14 gauge. Others recommend 18 gauge (1/16 in.). For 18-gauge sheet select a 1/16-in. filler rod. Both rod and torch tip are held at a 45° angle to the spot where the puddle is formed.

The torch is held by the mixing chamber and the filler rod is grasped 6 or 8 in. from the tip. Where it is held is not critical. It should be the point where it can be most comfortably maneuvered by the welder. You must, of course, hold it far enough from the tip so that the intense heat and radiation of the flame and molten metal will not burn through your gloves.

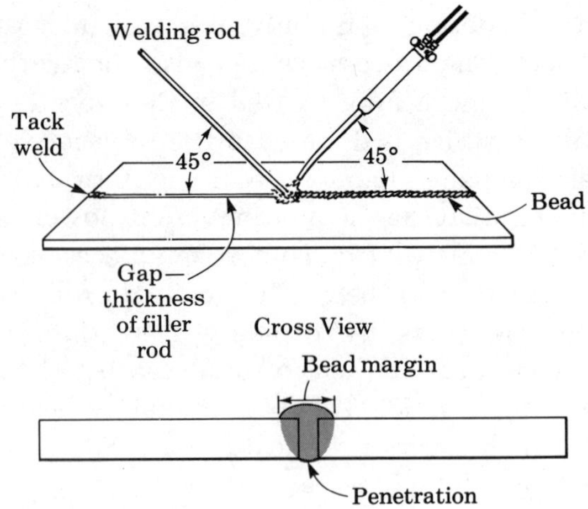

Figure 7-3. Holding the filler rod.

Importance of the Tip Angle

The 45° angle of the tip is important. If the tip is held too steeply, heat from the flame and puddle are reflected back into the torch tip, causing it to overheat. The heated torch tip will cause backfiring (small explosions at the tip).

On the other hand, if the torch is held at too shallow an angle, the puddle does not form as steadily. It is sort of blown along the joint with little or no penetration.

The 45° angle provides for good penetration, good puddle control, and good heat refraction to reduce tip temperature.

POSITION OF THE TORCH

The end of the inner cone of the neutral flame is held about 1/16 in. (about the diameter of the welding rod) from the surface of the metal to be welded. If you hold it closer, the molten metal in the puddle may possibly reach the tip. This will restrict the oxyacetylene gas flow from the torch.

If the inner cone is held farther than 1/16 in. from the puddle in fusion welding, the flame envelope will not be able to completely consume the oxygen in the surrounding atmosphere. The weld bead will trap the

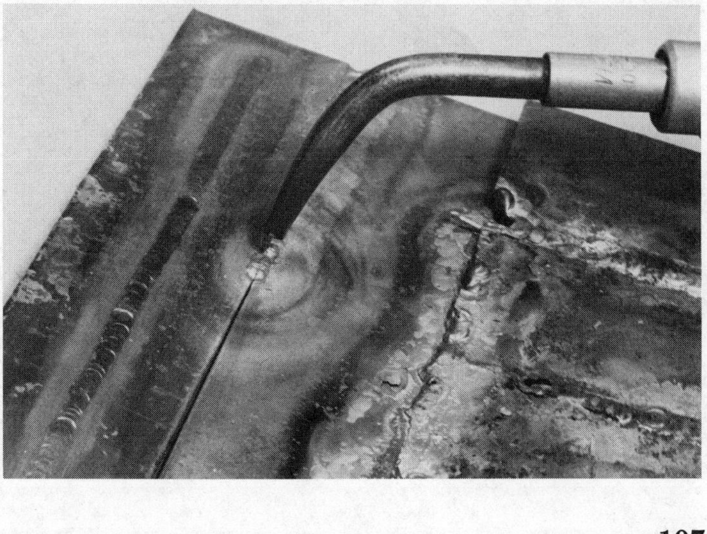

Figure 7-4. The tip of the white inner cone must be held about 1/16 in. from the base metal when doing fusion welding.

excess oxygen within itself. This will produce a weak, porous joint.

HEATING THE METAL

The torch flame is used to bring the surface temperature to the melting temperature. The welding rod is held *close* to, but not into, the flame. This heating of the rod makes the filler metal and the base metals easier to join. If the weld puddle is colder than the filler rod, the rod will stick to the metal instead of melting and flowing into it.

POSITION OF THE FILLER ROD

The filler metal rod is held at a 45° angle to the base metal. The filler rod is held directly in front of the inner flame cone to warm it. The rod is *not* melted by the flame. It is melted by dipping it rhythmically into the puddle, using the left hand to hold the rod and the right to hold the torch (for right-handed people).

Figure 7-5. The rod and the torch both should be held at a 45° angle to the base metal. In this picture the rod is held lower than it should be.

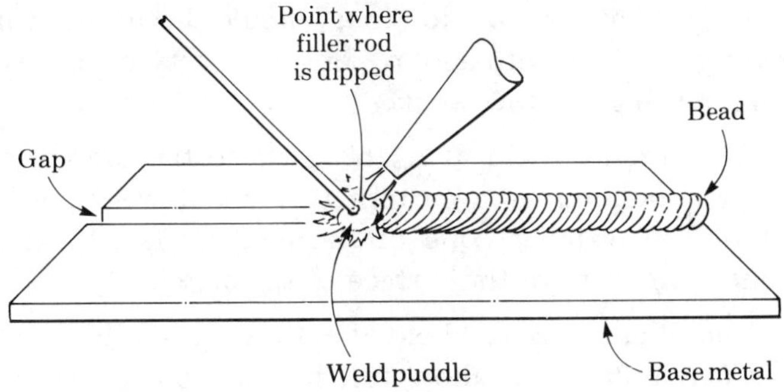

Figure 7-6. Positioning the torch.

When the filler rod is dipped into the puddle, it should touch the molten metal at about the outer one-third of the puddle's diameter, as shown in Figure 7-6. This control will allow the puddle to progress in a forward direction and not just pile up in one position.

HANDLING TORCH AND ROD TOGETHER

The rod and torch are manipulated as a team. There are several ways to do this to control the height of the bead, the width of the bead, and the penetration of the weld. These ways will be explained in the next chapter.

SUMMARY

1. Filler metal is necessary to build up a weld bead to fill the area between the parts and to add metal necessary to make a strong weld.

2. The filler metal must be as close as possible in composition to the base metal to achieve a strong fusion weld.

3. Filler rod specifications are set by the American Welding Society. Filler rods can be purchased in different diameters and carbon content to match required jobs.

4. Specialized rods are required for brazing, braze welding, and for nonferrous metal welding.

5. Rods should be stored in a cool, dry place to prevent rust. Rust, an oxidation of iron, will contaminate a weld.

6. Special high carbon rods are used for refacing, or hard facing. This procedure restores worn implements, giving them a hard coating to withstand wear.

7. Rods that are 30 in. long should have the non-welding end bent into a loop. This is a safety measure to avoid injuring a fellow worker.

8. The torch is held at a 45° angle to the base metal and manipulated in a pattern to control weld puddle agitation. The inner cone of the neutral flame is held about 1/16 in. from the surface of the metal.

9. The filler rod is also held at a 45° angle to the metal. It is manipulated to allow full mixture of the melting filler metal into the molten base metal in the puddle.

NEW WORDS AND TERMS

American Welding Society An organization that serves the interests of the welding profession.

Braze Welding A brazing process that also uses filler material of a lower melting point than the base metal. But where straight brazing filler is distributed by capillary attraction into a tight joint, braze welding lays a bead in grooves, spot welds, plug welds, and slot welds.

Brazing A process that uses the oxyacetylene welding torch to join metal by means of a filler that melts at a low heat and does not fuse or melt the base metal.

Filler Metal Material of the same basic composition as the base metal. It is added to the weld joint during fusion.

Hard Facing A system to build up worn surfaces to restore usefulness of implements and tools.

Silver Solder Another method of joining metals without fusion. It uses a filler material containing some silver. Soldering is usually worked under 800° F; over 800° F, it is called brazing or braze welding.

8 FORMING THE BEAD

A welder takes pride and satisfaction in his trade and his skill. This skill is demonstrated in the quality of the bead he lays down into each weld piece.

The quality of the bead depends upon several factors:

1. The amount of heat applied
2. The correct speed to carry the bead forward
3. The correct movement of the hand in forming the puddle
4. The proper angle of rod and torch
5. The proper manipulation of the welding rod

A bead is used to join metal in fusion welding (oxyacetylene and arc) and in brazing and braze welding. Forming a bead for arc welding and brazing will be discussed later. Here we are concerned with joining ferrous sheet or plate metal with a bead formed by fusion with an oxyacetylene torch.

SHEET AND PLATE METALS

In the welding trade, metals thinner than 1/8 in. are generally referred to as *sheet metal*. Metals thicker than 1/8 in. are known as *plate*.

The same technique is used for both sheet and plate, but the thicker plate will require beveling or possibly multiple welding passes. Sheet metal can be welded without beveling, and, if properly done, a single pass should be sufficient for a good weld.

PREPARATIONS FOR MAKING YOUR FIRST BEAD

Before actually making your first bead you must prepare your metal, open the valves, adjust your acetylene and oxygen flow, and light your torch. These items were all covered in detail in previous chapters. They are all very important, and it would be well to review them carefully before attempting your first bead.

Welding Materials

Your first practice beads should be made on sheet metal strips about 1/16 (16 gauge) in. thick. The pieces should be about 2 in. wide by 3 or 4 in. long. The small pieces are adequate for beginning practice and save expensive metal.

After the metal is selected, you need a welding rod. It should be approximately the same diameter as your metal. In this case, it should be 1/16 in. You may use either the gray-colored low carbon steel rod or the copper-coated low carbon steel rod.

You should bend one end of the welding rod in a loop to avoid accidentally poking a fellow worker or yourself with the sharp end.

Next, select a torch tip to match the metal thickness. For the 1/16-in. metal, a Victor 00 or equivalent size tip will provide a good flame. A larger tip would give too much heat, and instead of a puddle we might burn completely through the thin metal.

The tip must be clean to give the proper flame. If not, then clean the tip with the file and reamer as explained in Chapter 7.

Adjusting the Regulators

After the tip has been selected and tightened—hand tight—on the torch, the oxygen and acetylene valves are adjusted. Use this procedure:

1. Turn on the oxygen valve all the way. Adjust the line pressure to about 6 psi with a Victor torch (for this job). Lesser pressure is sometimes used for small jobs where small tips are used. Make sure that you follow all safety rules. Turn on the pressure slowly.

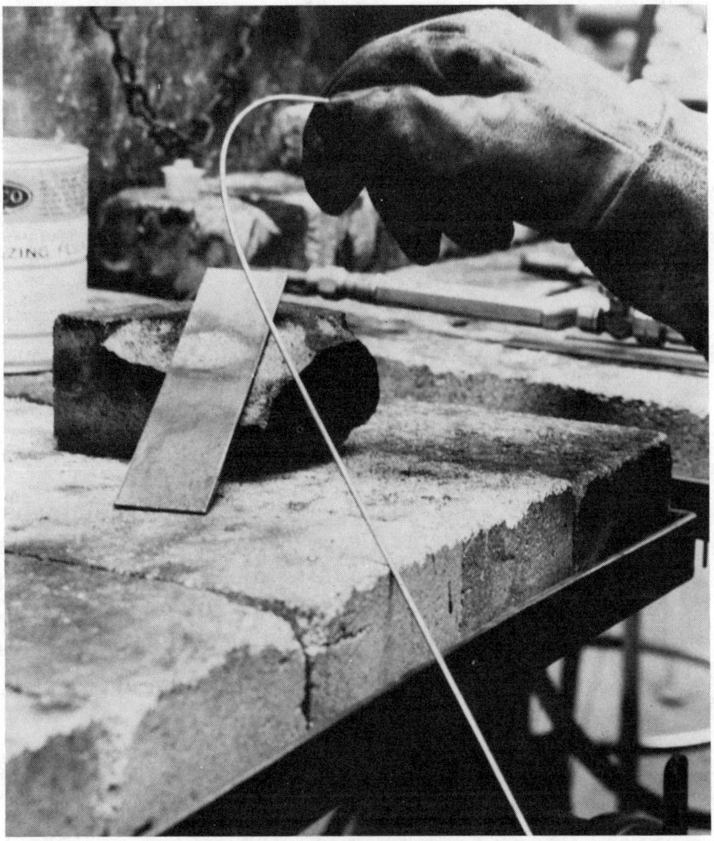

Figure 8-1. It is a good safety practice to bend one end of a welding rod to prevent poking a fellow worker or even yourself with the extended end.

Do not stand in front of the gauges as the pressure is turned on. Check for leaks. If you are using another brand of torch, consult the manufacturers recommended pressure for the size of torch tip.

2. Open the acetylene valve about one-half turn. Adjust the regulator to about 3 psi for this small job. Also check the cylinder pressure to make sure it is above 50 psi. If the gauge shows less than 50 psi in the tank, the cylinder must be taken out of service and replaced. If the pressure is less than 50 psi, acetone will be drawn into the regulator, lines, and mixing chamber and clog them.

Lighting the Torch

1. Open the needle valve on the mixing chamber to give a slight flow of acetylene. Ignite the acetylene with a striker. Hold the torch and striker at arm's

Figure 8-2. Acetylene tank pressure must always show more than 50 psi to prevent acetone from being drawn into the regulator, lines, and mixing chamber.

length away from your body and any flammable material. (The flame flares up when ignited.) Never use matches to ignite the torch. You might burn your fingers.

2. After ignition, open the needle valve until the acetylene begins to burn without producing a black, sooty flame.
3. Open the oxygen needle valve until the inner cone of the flame is blue and sharply focused to a single image. Be careful not to go beyond this focus by adding extra oxygen because this would produce an oxidizing flame instead of the neutral flame we need.

With the metal cut to the right size for practice, the right or correct welding rod selected, and the flame

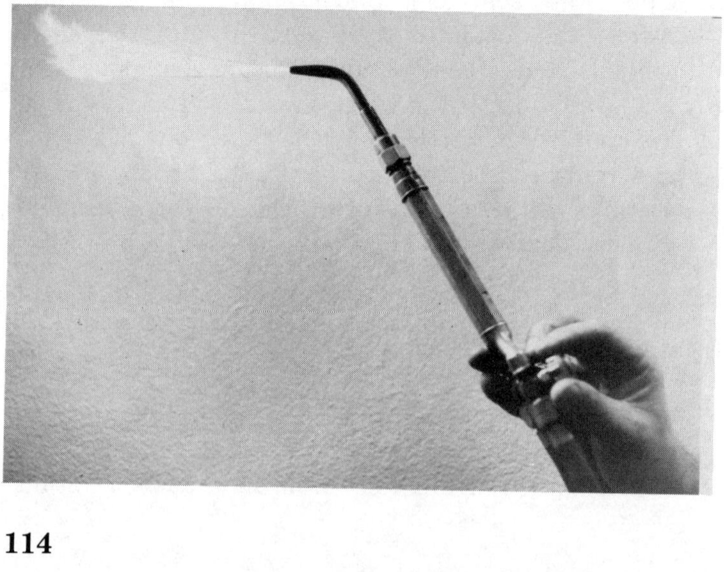

Figure 8-3. The torch is lighted and the acetylene is adjusted by opening the needle valve, as illustrated. The flame is further adjusted by opening the oxygen valve until the acetylene flame shown here becomes a neutral flame.

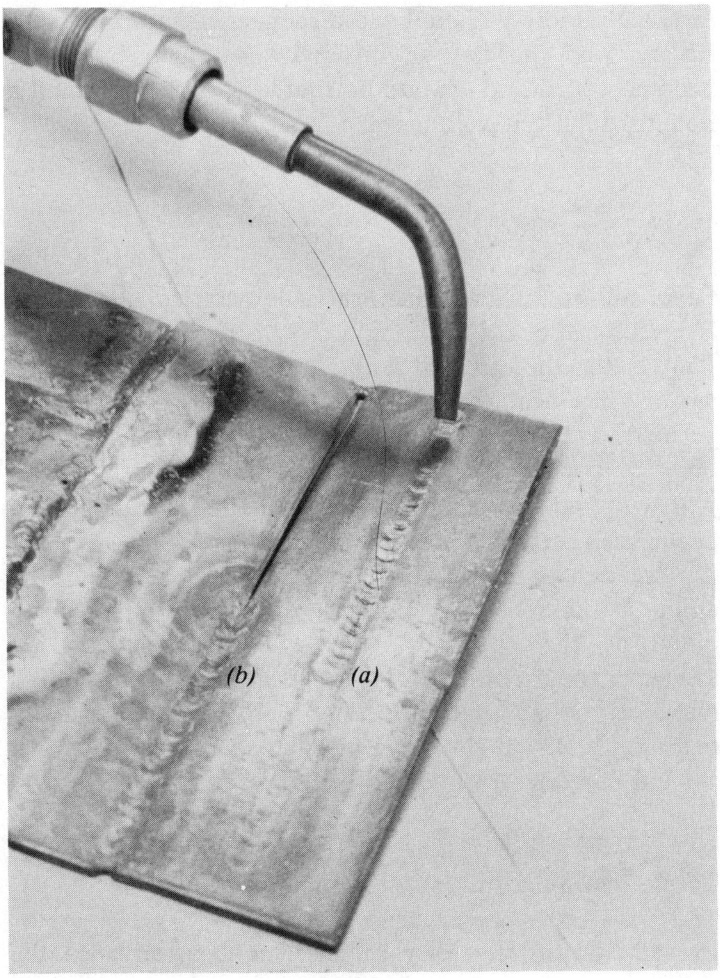

Figure 8-4. You can achieve proficiency in puddle carrying by doin a series of beads without filler rod before making an actual weld. (a) shows a bead made for practice without filler. (b) shows an actual weld.

adjusted, you are ready for your first assignment—how to "carry a puddle."

CARRYING THE PUDDLE

In building a weld bead we do not try to create one continuous stream of molten metal down the seam. Instead we build a series of overlapping puddles. Together these overlapping puddles build our bead.

This assignment is extremely important. It is the skill upon which all the welds you will ever make are built. Welding is more than just melting metal. You must be able to control the melted puddle and make it move down the seam.

The rate of movement is important. If you go too

fast, the puddle will cool and disappear. If you move too slowly, the puddle will absorb too much heat and become too liquid. A puddle that is too hot will melt a hole through the weld metal.

Torch Movement

Bead building begins with practice in control. The torch is held at about a 20° angle to the base metal with a slant in the direction of travel. For this initial practice, we are not using a welding rod. The tip of the torch is held 1/16 in. away from the surface. For this practice session use a constant torch angle (however, you will find that some variation in the torch angle can occur with different metal thicknesses).

Practice Session. The angle of the torch and the ability to keep the tip the same distance from the base metal as the puddle is moved down the seam are essential to a strong weld. At this point you should make dry runs (without flame) until you can easily maintain a 20° angle and a constant gap between torch tip and base metal as you carry the torch forward.

Torch Manipulation

In addition to the forward movement to advance the puddle, we must also use a side movement to create the proper molten pool of metal.

This is done through three common techniques. One is a series of simple overlapping circles [Figure 8-5*a*)]. These are similar to the writing practice we used to learn our letters in the first grade.

Another torch handling technique is to zig-zag through a succession of half moons or arcs [Figure 8-5(*b*)]. A variation of this technique is the leap frog method [Figure 8-5(*c*)]. This method advances a little and then returns about half-way before advancing again. This forms an overlap on each return.

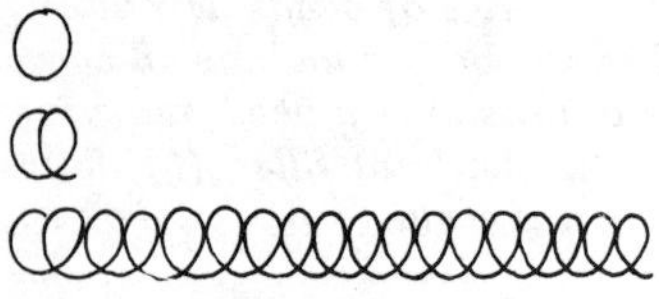

(a) Circular

(b) Zigzag

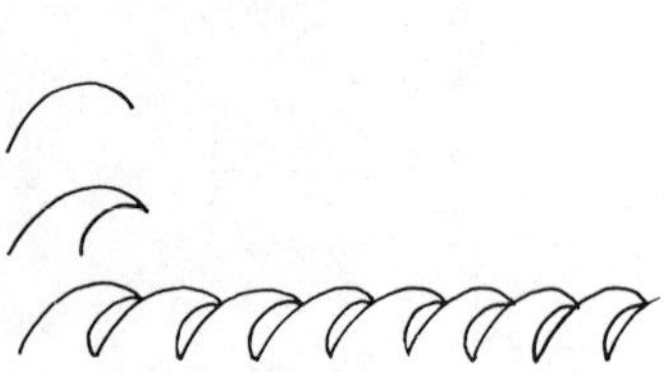

(c) Leap frog method

Figure 8-5. Torch-manipulation techniques.

Torch Dipping

In addition to moving forward and sideways, it may be necessary to slightly raise and lower the torch as well.

When you begin to oscillate or whip the torch from side to side, also practice raising the torch flame on the backward portion of the circle or half-circle. Then lower the torch flame as you move the torch to the front of the circle or half-circle. This dipping motion with the flame will cause the heat in the weld puddle to momentarily cool as the torch is slightly raised. This slight cooling will prevent the puddle from becoming uncontrollably liquid. However, we must not cool the mushy puddle too much or we will get inadequate penetration.

You can tell when the puddle is too cold because the filler rod will stick instead of melting. If the filler rod sticks, increase the puddle heat by dipping slower or not raising the torch so high. This small break during the torch movement is the key to allowing the weld puddle to take shape before the next layer is added to it.

In the next section, when you start adding filler rod to your practice bead, you must coordinate your rod movement with this torch dipping so the two work together. This will help you get a hot weld puddle without burning a hole in the metal.

This assignment has been for instruction in torch handling only. No welding rod was used. You are trying to achieve puddle control by torch manipulation. Continue practicing forming a bead until you can walk a good puddle across the full 3 or 4 in. length of the sample metal.

INSPECTING YOUR BEAD

After laying a bead by walking a puddle across the base metal, an inspection will show what you have done right or wrong. If the weld bead is formed with a series of overlapping ringlets, producing an even ripple along the top with the sides even, then your rate of forward movement and side manipulation was correct [Figure 8-6(*a*)] and your dipping motion was correct. If the puddles sag a lot but still have ringlets or ripple ridges, you can assume that your manipulation (circular or zig-zag motion) was satisfactory but your speed (rate of forward motion) was too slow [Figure 8-6(*b*)].

If your weld is thin or skimpy with few ringlets, you probably went too fast or you held the torch too high [Figure 8-6(*c*)].

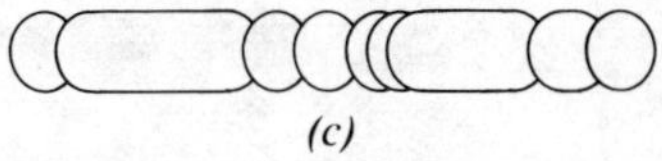

Figure 8-6. A correctly laid bead will have an even ripple and even sides (a). Sagging puddles, but still with ripples, indicates incorrect forward speed (b). A thin bead with few ringlets indicates too fast forward speed or torch held too high (c).

ADDING THE ROD

The next practice is adding filler rod to the molten weld puddle. In most welding beads filler metal is necessary. Flange (edge) and some corner welds can be made without filler, but the majority of joints require filler.

You must first get used to holding the rod and torch at a 45° angle to the base metal. Beginners often find this awkward, and it may take considerable practice until you can handle both properly.

Rod Manipulation

The rod is not held in the torch flame. It is held beside the flame so that it is heated, but not to the melting point. The rod is actually melted by dipping it into the molten puddle.

The awkward part comes from trying to manipulate the torch in a circle or zig-zag with one hand while trying to dip the rod in the puddle with the other. Many

Figure 8-7. Both torch and rod are held at a 45° angle to the base metal. This angle is important.

find this something like the old childhood trick of trying to rub your stomach with one hand while patting the top of your head with the other. You will find that both hands seem to want to do the same motion. It is, however, a technique that must be learned before going on to the next step. It is just a matter of intense practice.

JUDGING YOUR WORK

Eventually you will be able to make a satisfactory bead with the filler rod. The bead should have uniform ripples with a width about twice the diameter of the filler rod. Since our initial practice was with rod 1/16 in. in diameter, the ripples will be 1/8 in. wide, approximately.

The bead's height should be slightly above the surface of the base metal. If your hand is steady, the weld bead will have straight sides or bead margins, as they are called.

Figure 8-8. A butt weld bead hardens into a series of overlapping ringlets, thus producing a joint. A good bead will have an evenly rippled surface, with even height and width.

Figure 8-9. These are several samples of weld beads done by a beginner. Each bead shows the difficulty in trying to carry the weld puddle at the correct speed and width.

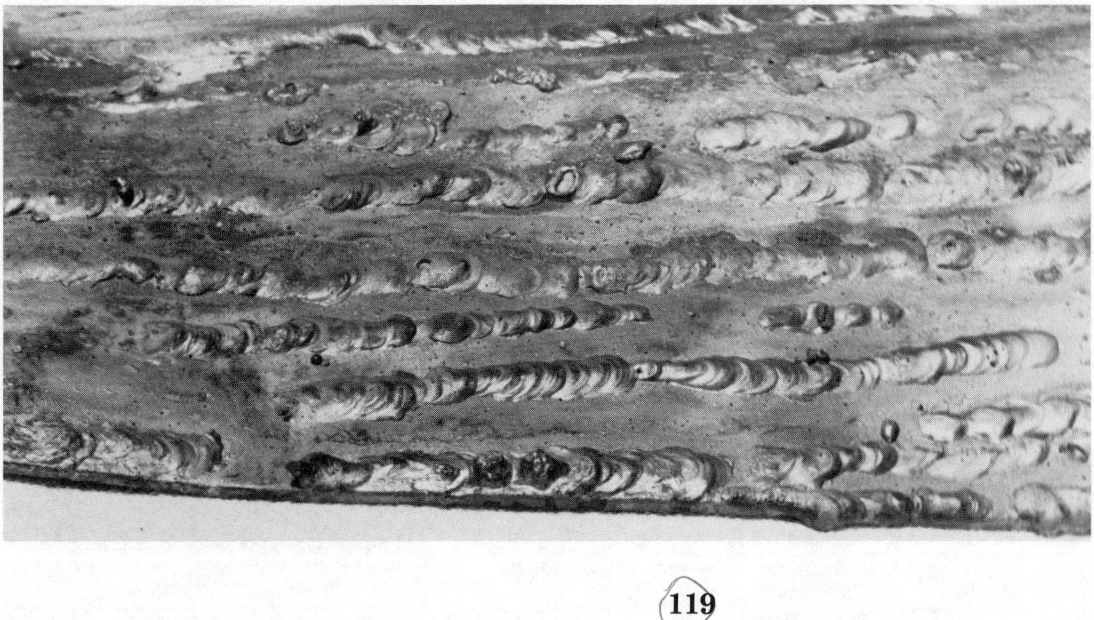

Sticking Rods

A common trouble beginners have is for the welding rod to stick in the puddle. This is caused by the rod being too cold. The rod should be heated, but not to the melting point. As mentioned earlier, the rod is actually melted by the heat of the puddle. Then all three pieces—the two pieces of base metal and the filler rod—will flow together.

If the torch flame is too hot, the puddle will not hold the proper ringlet shape and harden into a ripple bead. Instead, the puddle will sag and widen. You must judge the puddle heat correctly.

How to Judge Puddle Heat

Correct heat will give a bead with the right width, height, and penetration. Penetration means that the weld bead goes all the way to the bottom of the seam.

Full penetration is achieved by having a correct torch angle (45° to the base metal), proper torch heat

Figure 8-10. If the welding rod is too cold, it will stick in the puddle, as it has done here.

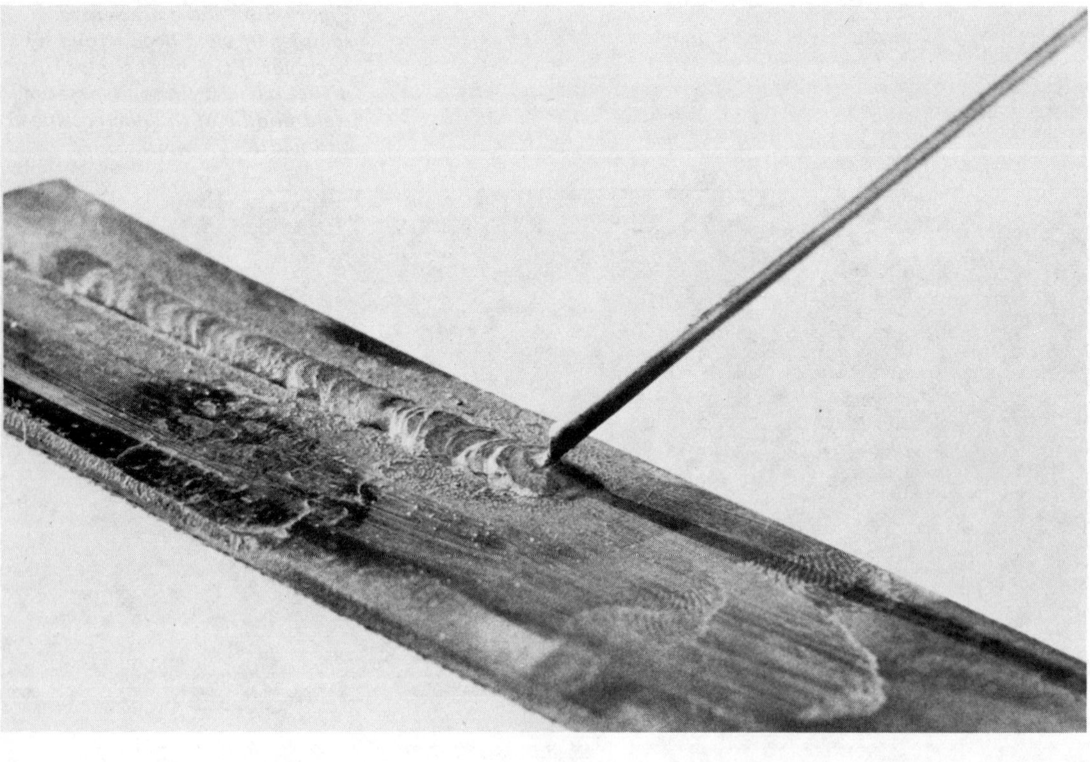

(just sufficient to melt the base, producing a puddle that does not sag or widen), the right torch rate (forward speed), and the correct gap.

Butt Weld Gap

The butt weld gap is necessary because metals expand as they are heated. We allow for the expansion by separating the two pieces of metal to be welded to produce a gap that varies in size from one end to the other. In Figure 8-11 you can see that the two pieces of metal are about twice as far apart at the left end as they are at the right end.

If a weld bead is laid down this gap, the heat of the torch as it begins welding the narrow end will cause metal expansion in the rest of the base metal. This expansion will cause the gap to close gradually as the torch proceeds down the bead. When you finish, the difference in the gap will have disappeared due to expansion and you will have an even joint between the two pieces.

Figure 8-11. Metal will expand as it is heated. To counteract this effect, you must allow a gap at the seam that is two or three times as wide at the end as it is at the starting point.

Controlling Warpage

If two pieces of metal are welded as described above, they will normally warp because of the difference in heat over the entire body of the two pieces being joined.

One way to control warpage is to clamp or lock the metal in a fixed position. Clamps or a clamping fixture called a jig are used to help prevent warpage. With this method the metal is placed in the jig and securely locked in place. The gap is set the width of the diameter of the welding rod. It is the same width from end to end and not angled as in the unclamped method above. The tight clamp of the jig prevents warpage. The welded joint is allowed to cool in the clamps or it is quenched in a quenching solution. This keeps warpage to a minimum.

Practice Session. Arrange two pieces of 1/16-in. metal with a gap ranging from 1/16 in. at the starting end to 1/8 in. at the opposite end. Practice making butt welds. Use the same torch and rod techniques you did in your puddle-carrying practice. Check the finished weld

Figure 8-12. This picture shows the buckling distortion caused by welding a flat bead without placing the base metal in clamps.

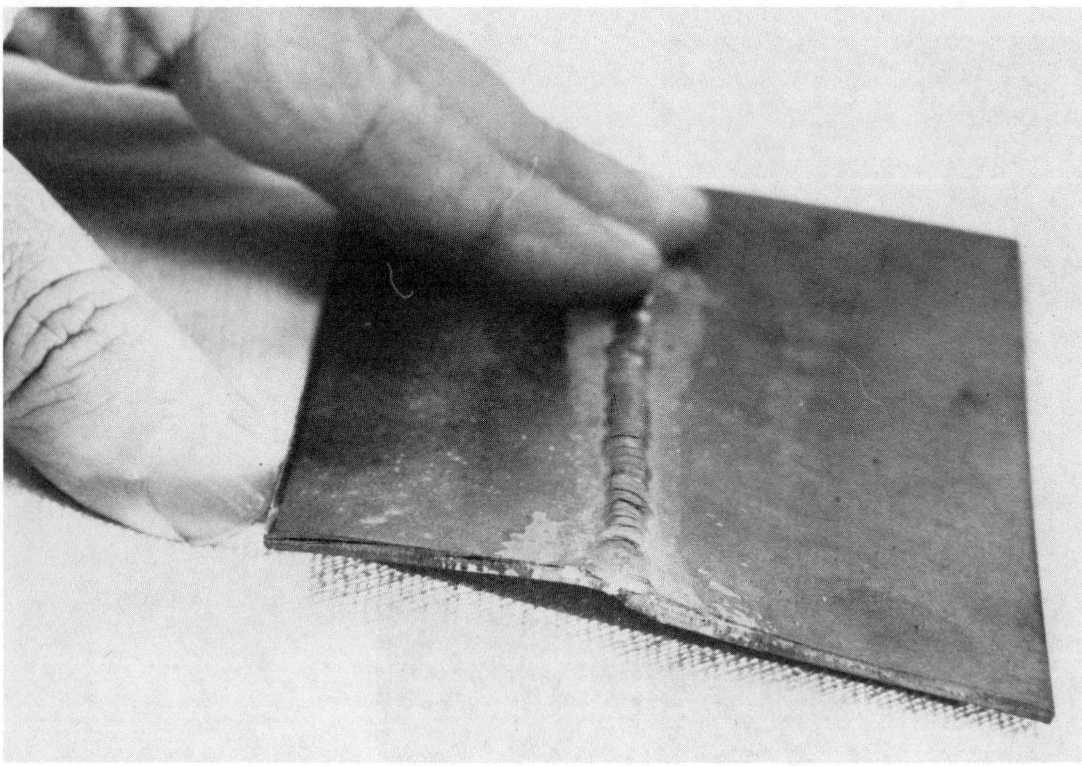

to observe the warpage. Next, using the same kind of metal, place the base metal in clamps or a jig and practice butt welds until you can produce beads with the proper ripple, height, width, and penetration.

THAT POPPING NOISE

Beginning welders often experience a lot of popping noises caused by small explosions at the torch tip. As discussed earlier in connection with the care and cleaning of torch tips, this popping is caused by a number of things. The noise can startle you and make you spoil a weld bead. When you have this trouble, check these possibilities:

1. Wrong size tip—too large
2. Incorrect regulator adjustments
3. Tip too close to the surface of the base metal
4. Wrong angle of the torch. Keep it at about a 20° angle
5. Dirty tip
6. Loose tip or poor seal between tip and mixing chamber

That popping noise is backfiring. Be sure and check the items listed and correct this condition. If the backfire is strong enough, it can travel inside the tip, lines, and into the cylinder tanks and cause the tank to explode. This is called a "flashback." One-way valves should be installed in the lines at the regulators or mixing chamber to stop this condition. If you continue welding with these small explosions, you may explode the welding hose in front of the one-way check valve.

TACK WELD

The use of a clamp or jig can help control warpage, but there are times and conditions where metal cannot be locked in this manner. In such cases we use a tack weld to help control warpage. The tack weld is a small spot weld used to hold the two pieces of base metal in position. After the two pieces are locked together by the tack weld, you can weld in between the tacks, covering about 1 in. at a time and stopping between

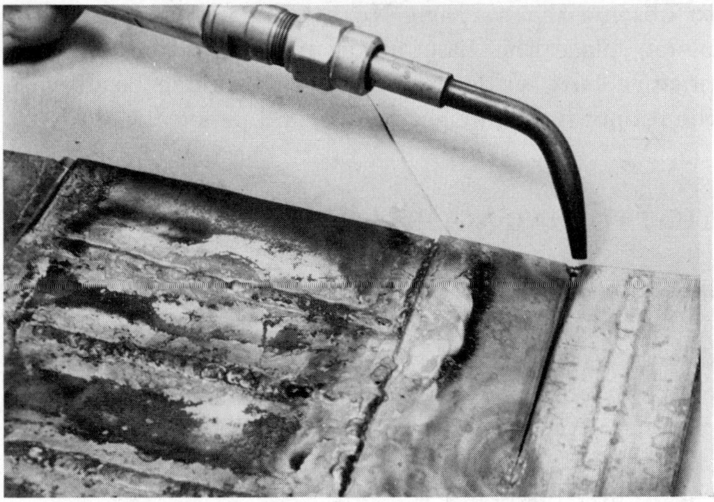

Figure 8-13. To help prevent distortion, this seam is tack welded at each end before the bead is laid down.

each weld to let the metal cool. This will keep warpage to a minimum.

OTHER JOINTS

This discussion has been restricted to the butt weld, using a flat bead. Once you have mastered the art of laying a flat bead through practice, you have acquired the skill to handle other types of joints used in fusion welding. Some of these joints require special techniques because of the position of the joints, but they are all based upon the ability to manipulate rod and torch to lay a proper bead.

The various joints used in fusion welds were covered in the chapter on preparing metal for welding. We will now consider the individual problems in making each kind of joint.

CORNER WELDS

The outside corner weld is first set up with two pieces of metal at a 90° angle. The two metal pieces will lay against each other with no gap.

The weld bead is laid on the outside corner of this joint. It forms a very smooth bead and can be done with no filler rod for metal thicknesses up to about 1/8 in. If the metal is thicker than 1/8 in., you will need to add filler rod to fill the joint.

Figure 8-14. These fillet welds are being done on a T joint. The torch is used to control the weld puddle as the filler rod is added into the weld puddle.

Another way to make a corner weld is to slightly overlap one piece of metal. This projection is melted to weld the two pieces together.

Begin with thinner sheet metal. Heat one end carefully while watching the color of each piece. When both pieces have the same color and the weld puddle is beginning to have a wet, shiny spot on it, angle the torch flame very steep to the weld seam—about 15° to the surface. This angle will help the pressure from the torch gases move the puddle along the weld seam.

For the thinner metals it will not be necessary to oscillate the torch because the weld puddle will be very fluid and will weld quickly along the seam. As metal thickness increases, you will find that some oscillation will be helpful in controlling the weld puddle.

Thicker metal welded as an outside corner weld will require filler rod. The addition of filler is done in the usual manner. Carry the puddle with torch oscillation, using a dipping motion to add the rod and control the weld puddle. This will help you make a strong, smooth bead.

FILLET BEAD

A fillet bead joins two pieces of metal at a 90° angle. It is used for making the lap weld and the inside corner weld. It slopes across the right angle of the base metals.

The easiest way to make a fillet weld is to first tack

Fillet Weld

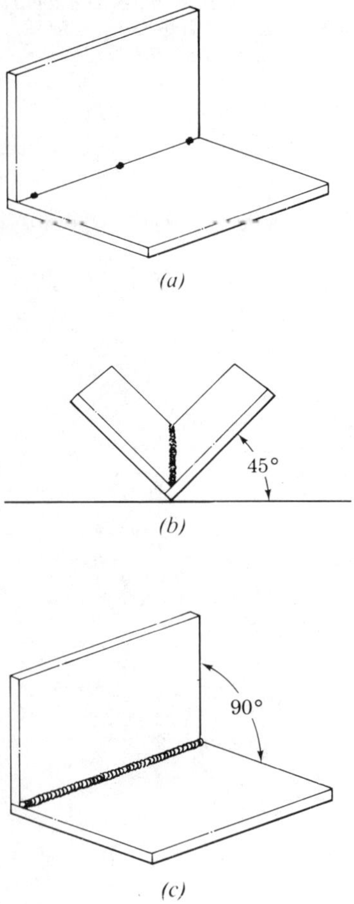

Figure 8-15.

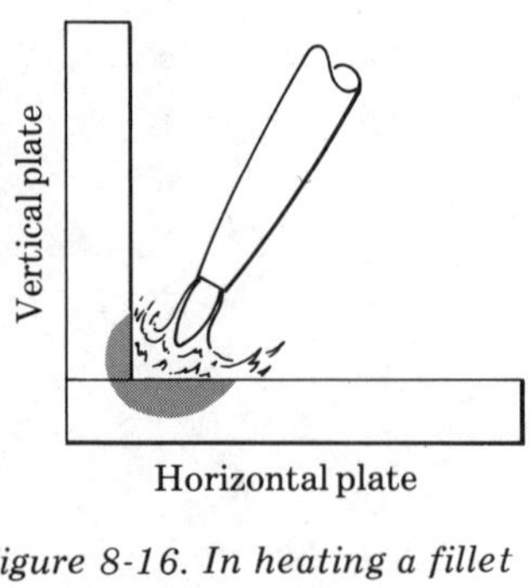

Figure 8-16. In heating a fillet weld, the flame is directed more to the horizontal plate than to the vertical plate.

weld the two pieces together to hold them in proper relationship to each other. Then turn the metal until the two pieces form a 45° angle to each other as shown in Figure 8-15. This creates a flat vee trough along which we can lay a bead. In this particular case, the running of the fillet bead is not much different from laying a flat bead in a vee groove [Figure 8-15(*b*)].

However, there are many cases where it is not practical or possible to take this easy way out. The fillet weld must be made with the members to be joined at right angles to each other [Figure 8-15(*c*)]. This type of weld is slightly more complicated than a straight flat bead.

FILLET WELDS ON T JOINTS

A fillet weld on a T joint (or an inside corner weld) requires more heat than the flat bead weld to keep the weld puddle fluid enough to control penetration. The penetration problem occurs at the bottom of the bead where the two metal pieces meet at a 90° angle.

Our problem here is to properly heat both the flat bottom piece of metal and the upright piece. The torch flame should not be aimed directly at the center of the inside corner. This would reflect heat back into the torch tip where overheating would cause a backfire. Also, if the torch flame were directed at this root where the two pieces join, the weld puddle would be unstable and would wander.

Because of these problems, we direct more of the flame to the bottom piece than to the top piece. Heat from the bottom metal will be reflected up and heat the upper piece sufficiently.

During this heating, we watch the color of the metal carefully and bring the temperature of both pieces together evenly. This can be done even though the torch is directed more to the bottom than to the top piece.

Adding Filler Rod to a Fillet Weld

When the metal approaches the lower critical temperature and the wet shiny spot is beginning to form on the metal, add the filler rod. As in the other fusion welds, the filler rod is melted by dipping it in the molten puddle. It is not melted by the flame.

The puddle will tend to stack up by forming a natural slope to the 90° angle. Care should be taken not to add too much filler rod or the puddle will bunch up at the bottom of the joint.

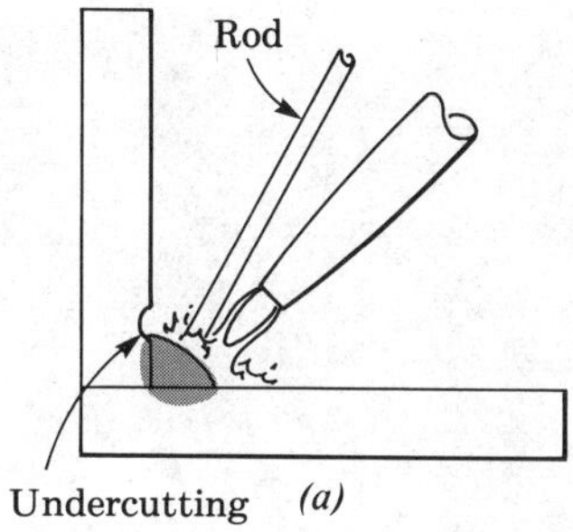

(a)

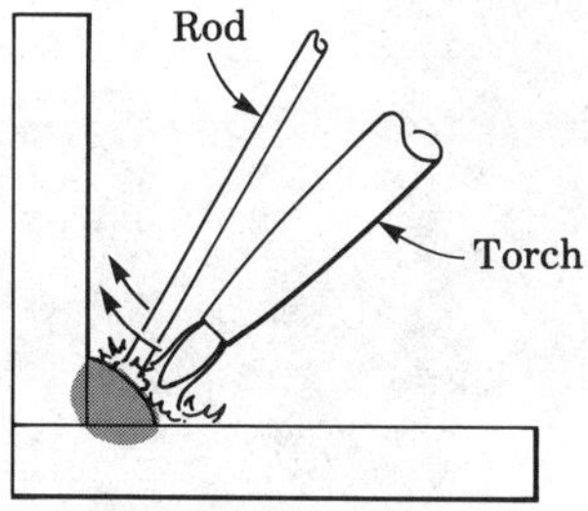

(b)

Figure 8-17. Undercutting is a depression above the bead where the flame has eaten out the metal and the molten pool has sagged down (a). This is corrected by a slight upward motion of rod and torch, which pushes molten steel into the depression (b).

Undercutting

If the torch and rod are not manipulated properly in making a fillet weld, undercutting may occur. An undercut occurs when the puddle does not have sufficient height and width. The fillet weld should have a weld bead that has width and height of about the same distance. In other words, the weld bead should extend up the upright member as far as it spreads out on the bottom member. The width is generally twice the diameter of the rod.

Figure 8-17(*a*) shows an example of undercutting. The torch has melted the upright member, but instead of solidifying in place the puddle has sagged, leaving an exposed cut or hollowed-out area just above the weld bead. This weakens the metal.

Preventing Undercutting

Undercutting can be prevented by manipulating the rod and torch in an upsweep of the weld puddle during the dipping motion of the rod into the puddle. You do this by slightly raising the flame and rod. This moves the weld puddle into the undercut area. The molten metal then freezes itself into the undercut area, forming a solid bead.

LAP WELD

Lap welds are joined with a fillet bead since the joint is also made of a 90° angle. You will find it easier because the top edge of the lap joint can be melted with the filler rod to form part of the weld bead.

Control of the puddle with torch and rod movement are the same as for the T joint. Be sure that there is enough filler rod to give a good uniform bond.

Figure 8-18. This picture shows a multiple pass fillet weld used for a T joint.

MULTIPLE PASSES

There will be times when a single bead will not fill the joint to the necessary height. It will then be necessary to make more passes to build the bead to the proper height. The thickness of the metal and the type of joint determines the need for multiple passes.

Sheet metal is thin enough to be welded with a single pass, if the welder uses the right diameter filler rod and welds at the correct speed. Metal plate, however, often needs multiple passes. The plate is prepared as explained under joints. This means beveling to a single vee, double vee, or either of the several joints.

The first weld bead we lay is called the *root pass*. This is the most important because it is the one that ensures full penetration and provides the foundation for the extra passes that follow. After the root pass is made, the joint is then filled to the desired height by adding other weld beads alongside the bead margin of the root pass. The stacking of the weld beads will then form the shape of an upside-down pyramid.

In Figure 8-19(*a*) the root pass is shown filling the bottom of the vee groove, including the small unbeveled base. This is full penetration. Figure 8-19(*b*) shows how a second pass is used to finish filling the groove on comparatively thin plate. Figure 8-19(*c*) shows how a succession of passes are built up for thick metal plates. Sometimes a succession of several passes are used even on thinner metals. This is done when it is desirable to prevent as much heat build-up as possible because of possible warpage or the need to prevent heat changes in the base metal.

Slag—oxidized substances combined with flux—will form on the surface of each bead. This must be removed by cleaning each bead before making the next pass. This removal will ensure that the succeeding beads fuse into the root pass and themselves create a strong bond.

Practice Session. The fillet weld is one of the two basic beads in all types of welding. It must be mastered before you can proceed. The fillet is used any time the bead must join metals that cannot be butt welded. It is used on the lap weld, inside and outside corner welds, and the T welds. The fillet bead, in all its numerous forms, must be practiced until you can achieve a satisfactory bead with all of them. This is a basic achievement that cannot be neglected. It is not difficult, but it does require practice to master.

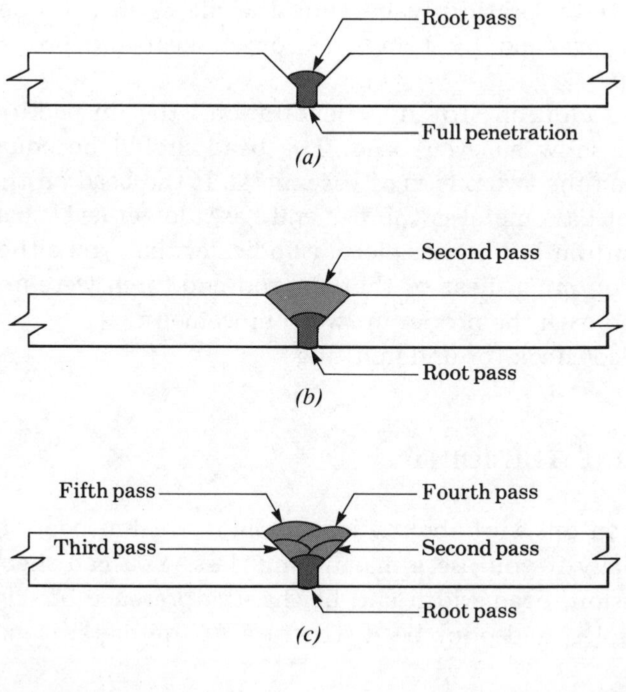

Figure 8-19. Multiple pass joints are made by firest depositing a root pass (a). This is followed by either another single pass to fill the vee groove (b) or by several extra passes, if necessary (c).

Figure 8-20. This picture shows the underneath side of a weld with poor penetration. Because of either too small a tip or failure to bevel thicker metal, the base metal was not melted all the way to the root.

FILLET WELD INSPECTION

Penetration is essential and is ensured (if plate is being welded) by previously beveling the pieces in a proper vee groove and by using the correct heat and proper size tip.

In addition to full penetration, the inspection should show an even bead. The bead should be equal between the two pieces of base metal. If the bead on the vertical base metal is thinner and has a lower level than the bottom horizontal piece, it indicates that you either used too much heat or that the rod and torch were not worked with the proper upsweep movement.

Also check for undercutting.

TESTING THE JOINT

You can tell a lot about a joint from a visual inspection, especially if you use a magnifying glass. You can check for fusion, bead width and height, the presence of slag, blowholes, and poor bead structure. If the metal is in a

Figure 8-21. Welds can be checked by destructive testing. The metal is clamped in a vise and bent with a hammer until it breaks. Such testing destroys the metal, but it is a good way for beginners to test the quality of their work.

position where you can see the bottom, you can also check for full penetration. This is possible only on butt welds. The only way we have to check penetration on fillet welds is by destructive testing or by the use of x-ray and sound (sonic) machines. The latter are available only in large industrial plants.

Destructive Testing

Destructive testing will show what is wrong with your weld. Unfortunately, it also destroys the weld. We cannot use it on every bead we make, naturally, but we should use it on our first practice fillet beads to find out if we are making them correctly. After we learn to make good beads, then we will use destructive testing for an occasional spot check.

One way to make a destructive test is to put the test metal in a vise and then bend it at the weld, using a heavy hammer (Figure 8-21). If the metal splits, as it has in this example, then check carefully to determine the cause.

Some of the causes are

1. Using too small a tip on the torch, thereby producing insufficient heat.
2. Moving too fast with rod and torch, preventing complete fusion.
3. Trying to weld thick metal without beveling.
4. Not cleaning the metal of oxides before welding.
5. Use of the wrong size of welding rod.
6. Use of the wrong alloy of welding rod.
7. Trying to weld incompatible metals. If the base materials are not of the same type, then it may be necessary to braze instead of fusion weld. Brazing will be covered in Chapter 11.

PIPE WELDS

Welding pipe is an important part of every welder's job. It looks difficult but is actually no more so than welding sheet metal and plate. We use the standard flat bead and fillet bead. If you have learned to do these properly, you will have no trouble doing pipe welds. The major difference is that we must use new methods of preparing the metal for welding.

There are three common methods of joining pipe. One of these is the pipe butt weld. It is no different from the flat butt weld except that we work with round stock. The two butt ends are beveled if they are more than 1/8 in. thick. Then they are brought together and clamped in place so they will not move out of line during the welding process.

Pipe 1/8 in. or less in thickness may be welded without beveling. Leave a gap between the two pieces equal to the diameter of the weld rod. Fasten the two pieces securely with clamps. Small pipe can be supported by placing it in the groove of an angle iron. Then tack the pieces of pipe together with three evenly spaced tack welds.

Begin welding the pipe by forming a bead at one of the lower third areas on the pipe's outer surface, formed by the three tack welds (Figure 8-23).

Complete this section and then roll the pipe so the second third can be conveniently welded. Complete the final third for the finished weld.

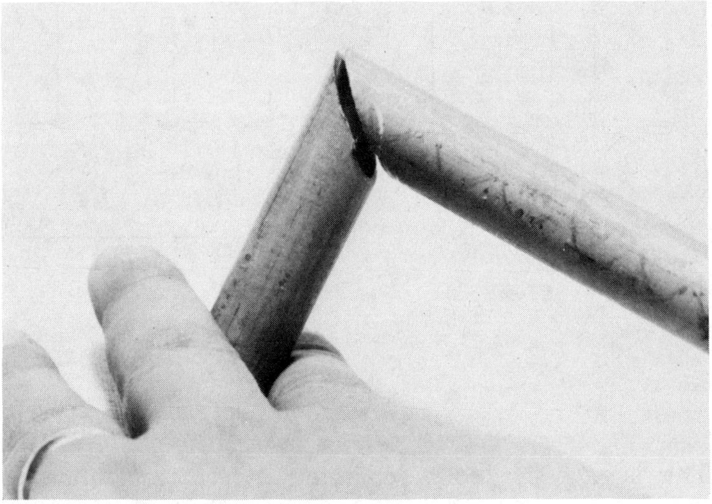

Figure 8-22. The prepration of the pipe joint is shown before the actual weld.

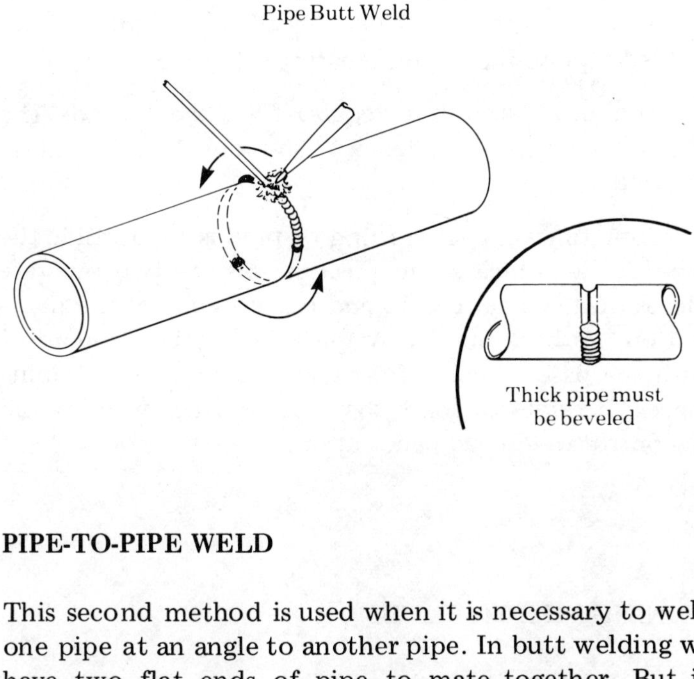

Figure 8-23. A pipe butt weld is made by making three evenly spaced tacks around the pipes (indicated by dots). This divides the pipe circumference in thirds. Begin welding on the lower third, and rotate the pipe as shown by the arrow to complete the second and then the final third.

PIPE-TO-PIPE WELD

This second method is used when it is necessary to weld one pipe at an angle to another pipe. In butt welding we have two flat ends of pipe to mate together. But in pipe-to-pipe welding we must fit the flat butt of one piece of pipe to the round surface of another.

We solve this problem by chamfering or cutting the butt of one pipe to fit the curving contour of the other. This can be more easily explained by referring to Figure 8-24(*a*). The first drawing shows what happens when the flat butt of the top pipe is welded to the bottom pipe. This would require filling the dotted line area with filler.

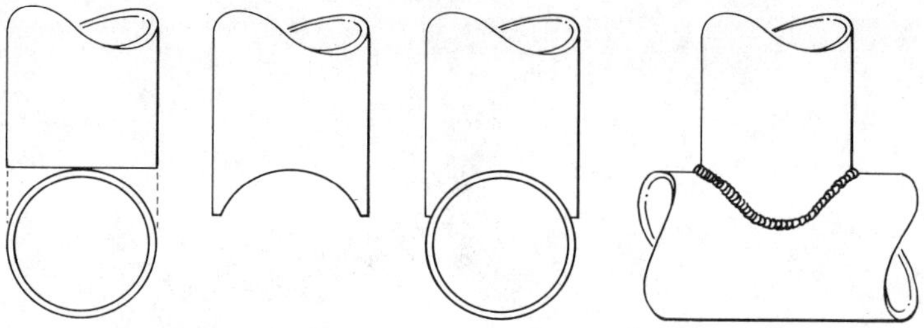

Figure 8-24.

This is excessive, so we chamfer—by grinding, filing, or torch cutting—the top pipe to give its butt the same curve as the pipe [Figure 8-24(*b*)]. It then fits snugly against the second pipe, as shown in Figure 8-24(*c*). Figure 8-24(*d*) shows a side view of the chamfered and welded pipe-to-pipe joint, creating a T weld.

The bead used is a regular fillet weld bead. The contour chamfer is called a "fish-mouth" or "saddles" by welders.

Sometimes a back-up ring or pipe is used inside the pipe to help locate the position of the two separate pieces during a butt weld and also to control the penetration. A back-up ring or pipe should be compatible with the base metal. It fits snugly inside the two joining pieces. It serves to space the pipes for tack welding and to control weld bead penetration.

JOINING PIPE TO PLATE

The third type of pipe weld joins pipe to a piece of sheet metal or plate. This also is done with a fillet weld. The bead is formed and continued around the entire pipe. Always overlap the bead at the end to prevent any gaps.

The three types of pipe welds are shown in Figure 8-25: (*a*) is a butt weld; (*b*) is a fish-mouth pipe-to-pipe weld; and (*c*) is a pipe-to-plate weld.

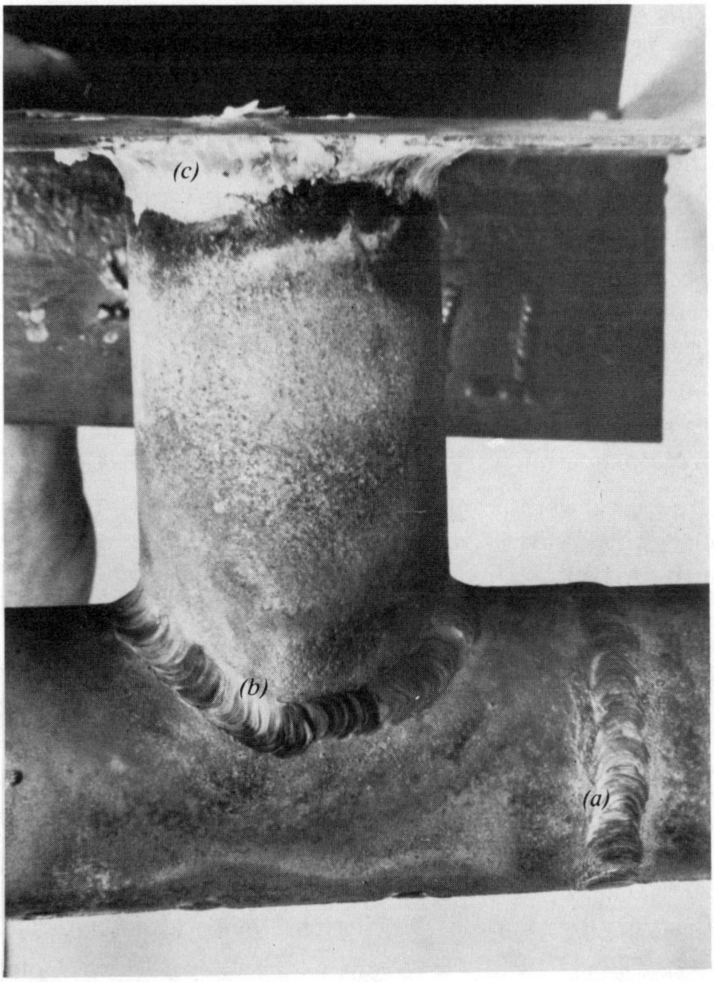

Figure 8-25. Three types of pipe weld are shown here. (a) is a standard pipe butt weld. (b) is a pipe-to-pipe weld. (c) is a pipe-to-plate weld.

SUMMARY

1. Forming a bead is the basis of welding and must be mastered through careful practice.

2. Bead forming begins with preparing your oxy-acetylene equipment and adjusting the torch to the proper flame.

3. Bead quality depends upon the right amount of heat, correct forward speed, movement of your hand in forming the puddle, proper angle of rod and torch, and correct manipulation of the filler rod.

4. Begin bead practice without a filler rod to learn how to form a puddle and carry it across the metal.

5. After you learn to carry a bead and can control the weld puddle at the proper rate and angle, you begin practicing adding filler rod.

6. The filler rod should be about the same diameter as the gap in a butt weld.

7. The filler rod is preheated but never placed inside the flame inner cone. It is melted by dipping into the weld puddle. The rod is added with a dipping motion that matches the weave motion of the torch, which forms the puddle.

8. Two basic beads are used in fusion welding. One is the flat bead. The second is the fillet. The fillet bead is used to join metal butting at a 90° angle.

9. Metal distortion is a problem. It can be controlled to some extent by tacking the joint and welding in small sections, permitting the bead to cool between welds.

10. Pipe welds require special preparation to match a flat butt end to a round surface. This is done by fish-mouthing or chamfering the pipe to a contour that will fit snugly with the round surface.

11. Undercutting is a problem with fillet welds. It can be controlled by an upsweep movement of rod and torch.

NEW WORDS AND TERMS

Carrying the Puddle Moving the weld bead by ripple stages.

Destructive Testing Bending a weld bead until it breaks to observe penetration, fusion, and interior defects.

Dipping A method of slightly raising the torch to control weld puddle heat.

Fillet Weld A weld bead used to connect metals at an angle.

Fish-Mouth A term given to the cut of a pipe butt to make it fit the rounded contour of another pipe.

Multiple Pass The use of more than one bead for a weld joint.

Pipe Butt Weld A joint for joining two pieces of pipe. The joint consists of welding the flat ends together. If the sidewalls of the pipe are more than 1/8 in. thick, the butts must be beveled before welding.

Pipe-to-Pipe Weld A pipe joint where the butt end of one pipe is welded to the contoured side of another pipe.

Pipe-to-Plate Weld Where the butt end of a pipe is welded to a flat sheet metal or plate.

Plate Flat metal over 1/8 in. thick.

Root Pass The first bead laid in a fillet weld or a multiple pass weld.

Sheet Metal Flat metal stock 1/8 in. thick or less.

EFFECT OF HEAT ON METAL

Now that you have made your first weld beads with an oxyacetylene torch, you may have noticed some peculiar things happening to your metal.

In Chapter 3 we discussed the effect of heat upon grain structure and what happens to metal when it is heated to the critical range. In this chapter we are concerned with the things that happen to metal during and as a result of welding.

A REVIEW OF METALLURGY

The hardness of ferrous (iron content) metal is increased by the addition of carbon. The hardness of some alloys can also be controlled by the way the metal is cooled. The rate of cooling affects the metal's grain size. Fine-grained metals are hard and large-grain metals are comparatively soft.

The normal grain size of iron and steel undergoes a change as their atoms shift when the metal is heated to 1325° F. At this point low carbon steel grains begin to grow smaller until they reach 1550°F when they again increase in size until the metal melts.

High carbon steel undergoes a sudden change to a smaller grain size at 1325° F, then starts to enlarge up to the melting point.

If these metals are heated above their critical temperatures—the point where the grain change occurs—and then are permitted to cool normally, the metal reverts to its original cold-state grain size and degree of hardness. If the metal is cooled rapidly, such as being plunged into cold water, it will retain the grain size it had at the time it was quenched.

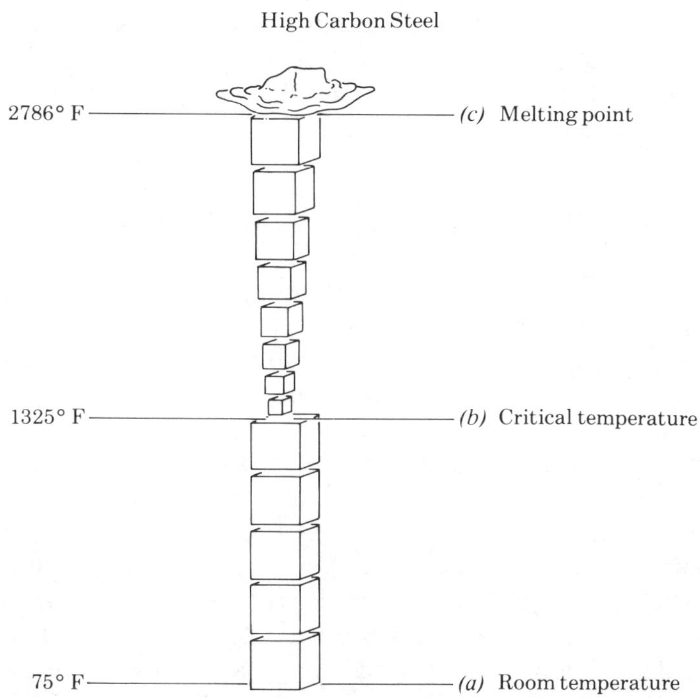

Figure 9-1. High carbon steel grains undergo a sudden change at the critical point (b) (1325° F) and begin enlarging again as the heat is increased to the melting point (c). Compare this sudden change to the gradual change of low carbon steel shown in Figure 3-2, page 32.

Figure 9-2. Quenching, which is plunging hot metal into a cooling solution, freezes the grain size.

This quick cooling is used in the heat treatment of metals to give them the desired hardness. Low carbon steel is difficult to heat treat. High carbon steel can be heat treated without difficulty.

DIFFERENT HEAT LEVELS DURING WELDING

The changing of the size of the metal crystallization (grains) is a very handy tool for altering the characteristics of steel to suit different purposes, but it is a genuine problem to welders. When forming a bead, there is danger of warpage due to heat along the weld seam. When a piece of metal is heated until the weld seam becomes liquid (molten), a wide range of temperatures in this one piece of steel are created. In the actual weld joint itself, the metal is molten with a temperature in excess of 2600°F. In the area next to the melting metal, the temperature will probably be 2200°F. As the heat radiates from the heated areas, adjoining areas are heated above the critical range of 1325° to 1550°F. In this range the metal hardens to its greatest extent, but farther to the edge the metal may be below the critical range with the grains at their normal size.

Thus we have in a single piece of metal a liquid state where the grains disappear, a critical range state of maximum hardness, and areas of comparative softness. This tremendous difference throughout the piece of metal causes pressures inside. These pressures or stresses do not disappear when the metal cools to an even room temperature.

SLOW AND FAST COOLING

Trouble comes from the fact that the heated metal cools at different rates within its total area. The molten bead will cool more slowly than the outside edges where the original temperature was less because the outside area is surrounded by cooler metal that absorbs and carries away more of the heat. The hot bead is surrounded by metal almost as hot as itself and this slows the cooling.

Slow cooling enlarges the crystalline grain structure of ferrous metals. This softens the metal and can remove any special heat treatment the metal received earlier. This means that welding can change parts of a piece of metal into an entirely different metal from the rest of the piece.

The changing of the grain structure, causing varying degrees of hardness in the metal, may set up stresses. Stress is pressure caused by unevenness in the metal and these stresses can lead to metal failures, producing strains and cracks. The crack in the Liberty Bell is a famous example of internal stress.

In addition to changing the grain structure and hardness of a metal, fusion welding can affect metal characteristics in other ways.

MECHANICAL PROPERTIES OF METAL

In discussing the changes in metal we have mentioned only two of its mechanical properties: hardness and stress. Other mechanical properties of metal are:

1. **Strain.** Caused by stresses in the metal.
2. **Elasticity.** The ability of a metal to return to its original shape after being stretched. Rubber is an example of an elastic substance.
3. **Tensile Strength.** This is what keeps a piece of metal from being pulled apart when loads are pulling it in opposite directions. A trailer hitch would need good tensile strength because the car is pulling it in one direction and the weight of the trailer is pulling it in another.
4. **Compressive Strength.** This is the ability of a metal to carry heavy loads. The wheels of a railroad car are an example. The entire weight of the car presses down on them as they turn.
5. **Torsional Strength.** This is the ability of a metal to twist without breaking. It differs from elasticity. An elastic metal will give some when pulled and then return to its original state. A metal with good torsional strength can be twisted, as is often seen in wrought iron decorations, without breaking or cracking (Figure 9-3).
6. **Fatigue Strength.** This is how long we can subject a metal to work before it weakens and breaks. Metal fatigue is an important factor in the breakdown of machines. It has also been the cause of aircraft crashes.

Figure 9-3. Good torsional strength permits a metal to be twisted without cracking or breaking, like this wrought iron table stand.

7. **Impact Strength.** This is the ability of a metal to withstand sudden pounding. If you strike a piece of metal with a sledge hammer and it does not break or bend, the metal has good impact strength.
8. **Ductility.** This is a combination of elasticity and torsional strength. It is the ability of a metal to stretch, twist, or bend without failing. You can consider it as a "not brittle" property.

PRESERVING METAL PROPERTIES

No metal has all the above-listed properties; for example, low carbon steel has greater ductility than high carbon steel. Low carbon steel, because of this greater

Figure 9-4. When a metal can be bent almost double, as here, it has good ductility.

ductility, will bend under the impact of a too-heavy load when a high carbon steel, given the same impact, would crack.

If a metal part of high ductility is welded, the ductile characteristic could be removed by an uneven heating and cooling. This could render the part unsuited for its original use. The same is true of other metal characteristics. Spring steel can lose its elasticity. A metal with high impact might be softened, and so on.

It is important that the welder preserve these original characteristics as best he can through proper heat control. One way to do this is by *skip welding.* Instead of laying one long bead, make a series of short ones, skipping space in between, then fill in the skipped places later.

Figure 9-5. A "rosebud" tip can be used for preheating metal. The large tip spreads flame over a larger area, raising metal temperature quickly.

AVOIDING METAL CHARACTERISTIC CHANGES

It must be emphasized that many of the metal characteristic changes are not caused by heating the metal, but because it is heated *unevenly*. This trouble can sometimes be avoided by *preheating* the metal. This means using the oxyacetylene torch to bring the entire metal up to a heat that cuts down so much difference between the different parts (Figure 9-5).

In one technique a second operator moves ahead of the first and uses the oxyacetylene torch to bring the metal to a predetermined heat before the second welder makes the actual bead.

REMOVING STRESS

Stresses, strains, and incorrect grain patterns can also be relieved by heat treatment after the weld is made. In one technique, the entire metal is heated in an annealing oven, soaked for a suitable time at an even temperature, and then cooled at a selected rate to restore the former heat-treat condition.

Stresses can be relieved by *peening*. Peening means to lightly hammer the metal with a ball peen hammer. Most authorities do not recommend peening by a beginner except as a last resort because it is very easy to set up new stresses by over-hammering.

DISTORTION

Clamps and jigs have already been mentioned as a means of preventing metal distortion or warpage caused by uneven heating. Sometimes chill plates are placed under the metal to absorb heat and lessen distortion. Chill plates are usually copper because copper heats rapidly.

Vee groove welds are especially subject to distortion because the filler at the top of the groove is thicker than at the root of the joint. This results in greater contraction at the top. This can pull a vee grooved butt joint out of line. For this reason vee-grooved butt welds are always clamped before they are welded.

Figure 9-6. This warpage was caused by not placing the metal in clamps during the welding process.

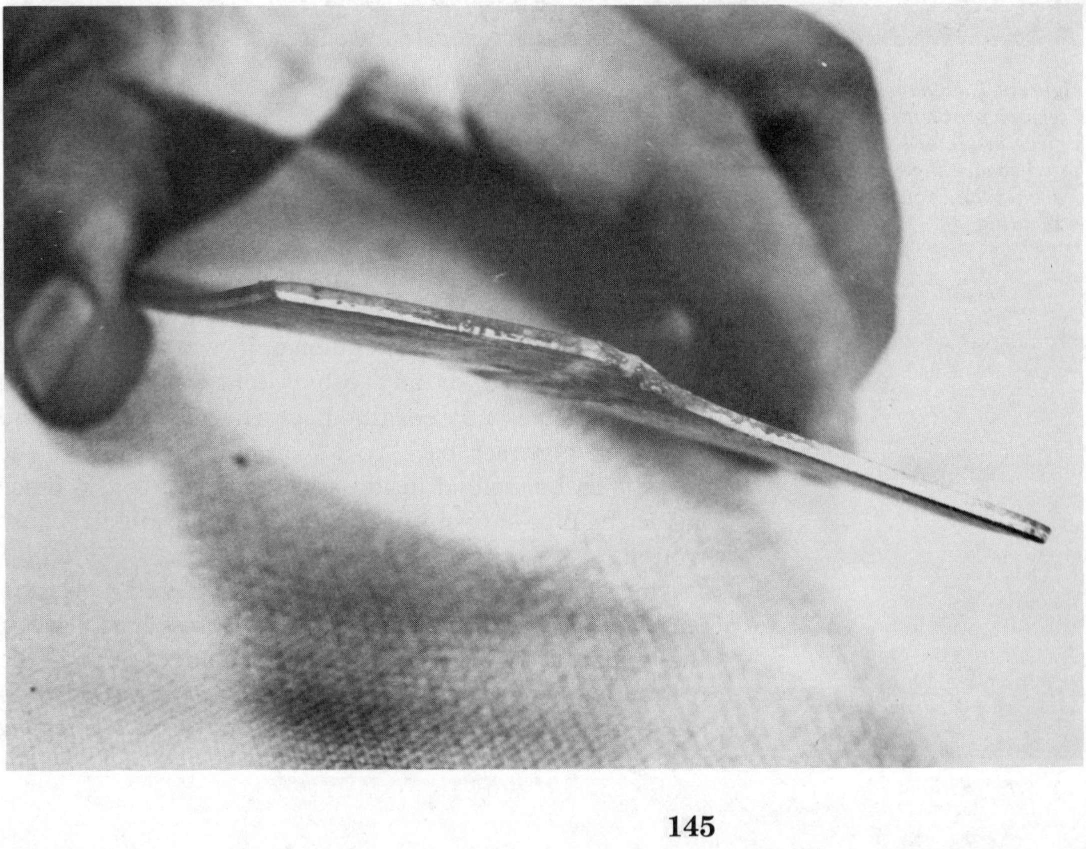

Figure 9-7. Expansion becomes a serious problem when welding a piece inside of a circle like here because the ring or iron prevents the center bar from expanding.

EXPANSION DIFFICULTIES

Metal expands when heated and contracts as it cools. This expansion is in all directions. If a rectangular piece of metal is heated, it will expand up and down, to both sides, and to both ends.

This is of little consequence if we are making a simple butt weld or any other joint where the piece being welded is not confined. Open metal will expand and then contract back to its original size, but if the piece is to be welded inside a solid object, as the brace inside the pipe shown in Figure 9-7, expansion can cause trouble.

In such a case the metal is not free to expand lengthwise. The linear expansion will press hard against the pipe. If the expanding iron cannot bend the restraining pipe, then it may buckle at the softer weld joints or in the center if the expansion pressure is strong enough.

Extreme care must be taken when doing an inside weld of this type. Chill plates can be used under the weld. Skip welds will also cut down on heat.

CASEHARDENING

In the heat treatment process steel is hardened by heating it above the critical point and then quenching the hot metal while the grain structure is at its smallest point. This method works well with medium and high carbon steel. Low carbon steel is difficult to heat treat; however, the addition of carbon to low carbon steel makes it suitable for hardening by heat treatment. This addition of carbon to low carbon steel is the basis of a process called casehardening. Casehardening means to harden only the "case" or outside of a metal part. This leaves the inside comparatively soft, which was a swordsmith's trick in centuries gone by. Warriors needed a hard cutting edge, but if the entire blade were hardened, it became brittle and would shatter in battle, so the swordsmiths developed the technique of leaving the inside of the blade comparatively soft while hardening the outer case.

Today casehardening is used for such industrial purposes as hardening gears. Gear metal must be ductile so that it can be easily shaped by machines, but a soft, ductile metal would wear out quickly. Through casehardening, soft metal can be used to fashion the gear and then the outer surface can be hardened to withstand wear.

How Metal is Casehardened

Hot steel and iron will absorb carbon, so the gear teeth, or whatever part is desired to be casehardened, is exposed to carbon or a carbon flame for a required length of time. This time will vary according to the type of steel, the temperature of the metal, and the kind of carbon. Generally speaking, sufficient carbon will be absorbed in an hour to harden the outer case of the part.

In casehardening the metal is not melted. It is heated and the carbon is absorbed into the hot metal to about 1/64 of an inch, but it can be deepened by lengthening the time the hot metal is exposed to the carbon.

After the case of the part has been carbonized, the part is then heated to its critical temperature and quenched. This will harden the outside while leaving the uncarbonized interior metal still ductile. The hardened shell provides a good welding surface.

SUMMARY

1. Heat causes drastic interior changes in the grain structure of ferrous metals. At 1325°F the grain diminishes abruptly in high carbon steel and begins to grow smaller in low carbon steel.

2. Heat also causes exterior changes that can be observed, including distortion and warpage.

3. Exterior changes are caused by the wide ranges of heat a piece of metal is subjected to during the welding process. Temperatures may range from the molten state in the puddle to room temperature at the opposite end.

4. Differences in temperature can cause varying physical changes in the grain structure which can set up stresses in the metal. Pressure of the stresses causes distortion and warpage.

5. Metals have a number of mechanical properties. They include hardness, elasticity, tensile strength, compressive strength, torsional strength, fatigue strength, impact strength, and ductility.

6. One problem of the welder is to preserve these characteristics and strengths in the weld.

NEW WORDS AND TERMS

Casehardening A method of absorbing carbon into the outer case of a metal part to make it subject to heat treat hardening.

Chill Plate A piece of metal, usually copper, placed under a weld joint to absorb heat and reduce the chance of warpage.

Peening A method of relieving stress in metal by light hammering.

Skip Welding A series of short welds made to reduce heat in a heat-sensitive metal. The spaces between the short skip welds are filled in on a second pass to make one complete long weld.

OUT-OF-POSITION WELDING

It is not always possible to sit down at a stool in a welding booth, clamp the metal flat on a table, support it with back-up blocks, and make the weld in a comparatively easy horizontal bead. Often you may have to weld out of position. This means working on projects where you find them, in any kind of position. Sometimes it is necessary to stoop, bend, squat, lie sideways or even on your back where there is danger of molten metal dripping down in your face.

The overhead weld is a challenge, but one that can be overcome by proper welding techniques. The basic problem in welding directly overhead is that we are subject to one of the oldest physical laws—gravity. We must form a molten puddle directly overhead and keep this puddle from either dripping or sagging enough to prevent complete fusion, thereby producing a poor weld joint.

We also face challenges in making a vertical weld and a horizontal weld on a vertical sheet or plate. You were introduced to vertical weld problems when making your first fillet welds. There was the problem of undercutting and the tendency of the puddle to sag. This you overcame by a slight upward sweep of rod and torch, which pushed molten metal from the puddle into the undercut depression left by the sagging metal.

In a similar manner we have techniques that will permit us to overcome the problems of out-of-position welding. In the fillet weld we had an advantage in that the vertical weld was only one-half weld bead high and it had support from the horizontal half of the bead.

A true vertical weld means to lay a bead on a vertical piece of metal by carrying the bead upward in a vertical position. We work from the bottom to the top of the joint.

Working a droplet of molten metal upward when it naturally wants to drip downward is something like pushing a bicycle up a very steep hill. It wants to roll back on you. If you move too slowly, the bicycle is harder to push. If you lose your balance, it falls.

Carrying your weld puddle straight up is a somewhat similar experience. If you do not move the puddle fast enough, it will become too fluid and will fall back. You overcome this problem by developing skill in balancing your puddle control. It is not difficult, but it does require practice to master.

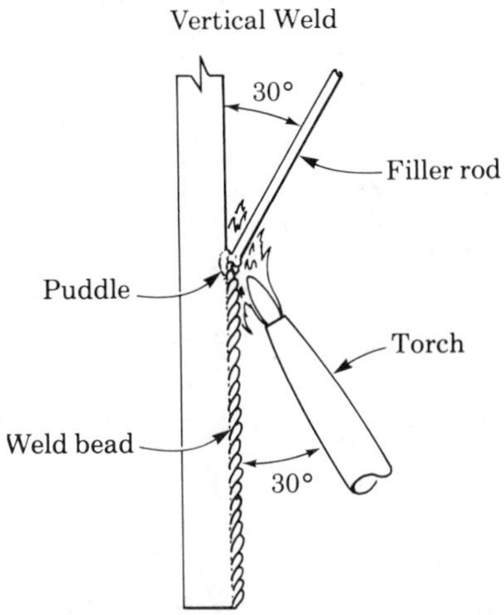

Figure 10-1.

SETTING UP THE VERTICAL WELD

The gap space for the vertical position weld joint is the same as for the downhand (that is, flat horizontal) weld joint. The joint may be tack welded in several places to hold it in position, or it can be clamped into a fixture that will secure both the spacing and help prevent warpage.

You will also use the same neutral torch flame. The torch and rod angles will be different. These were 45° in the downhand position. On the vertical weld we will change the angle to about 30°. The reason for this is that we want the pressure of the torch flame to push upward at a steeper angle to help us hold the molten puddle from sagging as we work our bead upward. Uphill welding is basically controlled by the balance and manipulation of the weld puddle.

UPHILL WELD PROCEDURE

The metal to be welded is heated until the puddle begins to glow and the small weld puddle shines like a wet spot. This is clearly visible through your filter lens. This is the signal that it is time to begin forming your bead.

Add the filler rod with the same technique you used with the flat horizontal weld, that is, dipping it into the molten puddle as filler is needed. At the same time you must work the torch with the same motions you learned in forming a bead: the circular, half-moon or arc, or zigzag motion.

Figure 10-2. This vertical up weld is held in position by a clamping jig. The torch and rod angles are shown at about 45° to the metal surface.

Figure 10-3. The weld puddle is heated with the torch tip until it begins to glow, forming a wet shiny spot.

While the rod is being added, it is most important to raise and lower the torch and filler rod together in a rhythm coordinated with each other. The raising and lowering of the torch controls the fluidity of the puddle. This prevents the puddle from becoming too fluid and sagging down. Each time the torch is raised just slightly, the molten pool is allowed to freeze or harden, which prevents the weld puddle from becoming too fluid and dripping, as would happen if each movement of torch and rod were not separated by this small rest break.

The entire secret of uphill welding is contained in this coordinated action:

1. Melt the metal to form a puddle.
2. Dip the filler rod into the puddle to add required metal.
3. Lift rod and torch momentarily to let the puddle harden below the sag point before dipping the torch back down as you move upward. The hardened ridge

Figure 10-4. The torch is being raised and lowered as the filler rod is dipped into the weld puddle. The control of the weld puddle is maintained in this manner.

that results provides a "shelf" upon which the next puddle can rest.

PRACTICE IS REQUIRED

Although a description of a vertical butt weld makes the procedure sound simple, it requires considerable practice to master. You may find it difficult at first to find the correct rate of puddle movement. Also, beginners experience some trouble in determining the correct speed for rod and torch manipulation necessary to freeze the puddle at the correct moment. As long as the principle of the uphill weld is understood, the rest will come with practice and experience.

One of the most common faults at the beginning is trying to go too fast; consequently, the weld puddle will be too cold. A cold weld puddle will cause the rod to stick when you dip it into the puddle. This is one mistake that is easy to recognize. If the rod sticks, your puddle is too cold. You lifted the torch too high or hesitated too long before lowering it again. The puddle must be hot enough to melt the rod.

BACKHAND TECHNIQUE

Up to this point we have used only one position of rod and torch. This is the forehand method. The rod is held in *front* of the torch and dipped into the *front* of the weld puddle. This position of rod and torch provides adequate penetration and weld bead formation. There is, however, another position for holding rod and torch—the backhand method. It is often used because it allows deeper penetration with about the same forward rate of welding speed. It is also useful in both vertical up welds as well as vertical down welds.

How the Backhand Weld Works

Figure 10-5. A backhand weld method is being used to go from left to right on this horizontal weld. Notice that the torch tip is pointing in the opposite direction of the weld bead travel.

There are two major differences between the forehand and the backhand methods. In the forehand method the

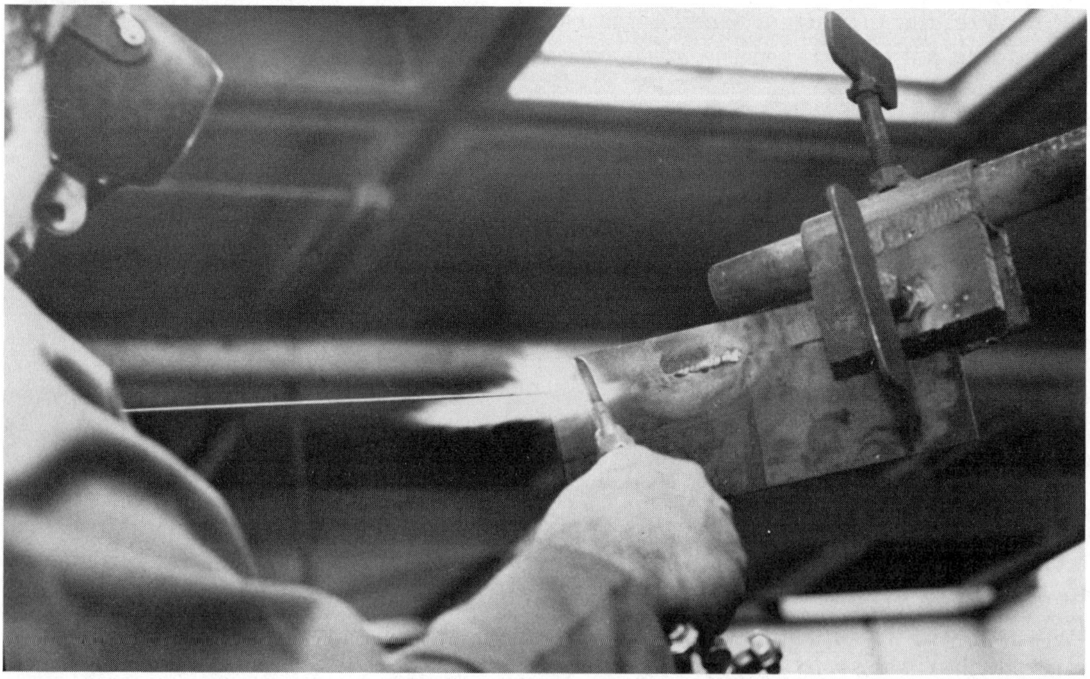

torch points in the direction of weld travel and the rod is dipped into the front of the puddle. In the backhand method the torch points in the opposite direction to the weld travel. The welding rod is dipped *behind* the torch: that is, to the rear of the puddle (Figure 10-6).

Both torch and rod are held at a 45° angle to the weld metal. The puddle is formed by the flame and the rod added to the rear of the puddle to form the weld bead.

This method results in a delay of heat transfer to the weld joint. In general, it provides for a less fluid weld puddle, and therefore it is a more convenient technique to use for vertical welds and pipe welds.

The appearance of the finished weld bead will be a little different also. Since the puddle is less fluid in the backhand method, the ripples will appear heavier and have a less polished look. The underside of the weld should look the same as in the flat weld.

You may experience some difficulty when you attempt your first backhand weld because the puddle is not as hot as you have been used to with the forehand method. This may cause you to stick your rod in the puddle. If your rod does stick, work your bead slower

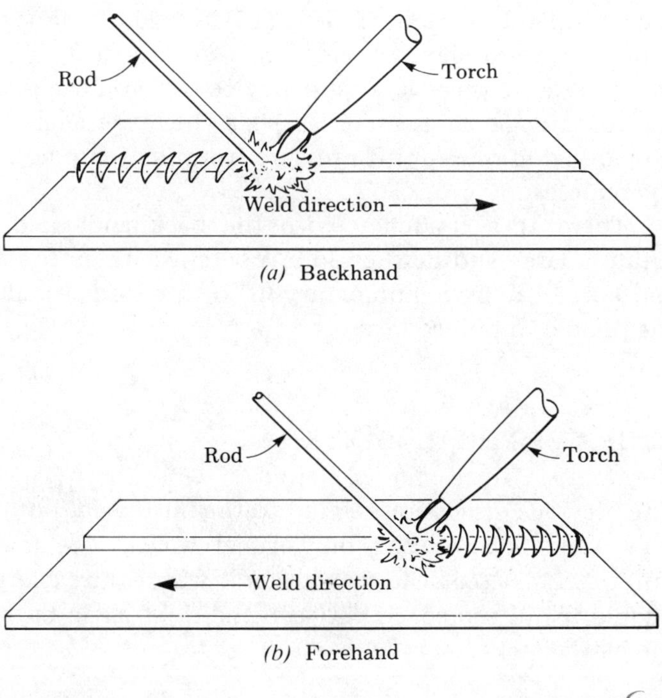

Figure 10-6. Differences between the forehand and backhand vertical welds.

and allow the weld puddle to stay molten. Use the side-to-side movement to control your weld bead, and add filler rod at the correct spot each time. Some practice with the backhand method in a flat position will give you the experience and confidence to work this technique in the vertical weld position.

VERTICAL DOWN WELD

In the previous exercise we completed a vertical weld by starting at the bottom and working our way upward. In the vertical down weld we start at the top and work down. The vertical down weld is often recommended for sheet metal because penetration is less than with the up weld method.

Again, we tack weld the two pieces with the proper gap before beginning the vertical down weld butt joint. The torch tip is pointed upward at a 10° to 15° angle. The filler rod is pointed downward at the same angle. This makes the torch point in the opposite direction to our line of bead travel.

Begin the weld bead by melting the tack weld at the top and applying filler rod to the puddle. The natural tendency of the puddle will be to sag down, but you can control the puddle by the pressure of the torch gases. In the backhand method the low (15°) angle of the torch and its upward slant provides this uplift effect against the puddle. If there is a tendency to sag, you can lower the torch angle to as much as 10°. The lower angle will provide additional uplift pressure from the torch against the puddle.

There is a tendency with the backhand weld to create a large and fluid puddle. You must watch for this and correct it by a momentary lift of the torch to allow the puddle to reduce heat.

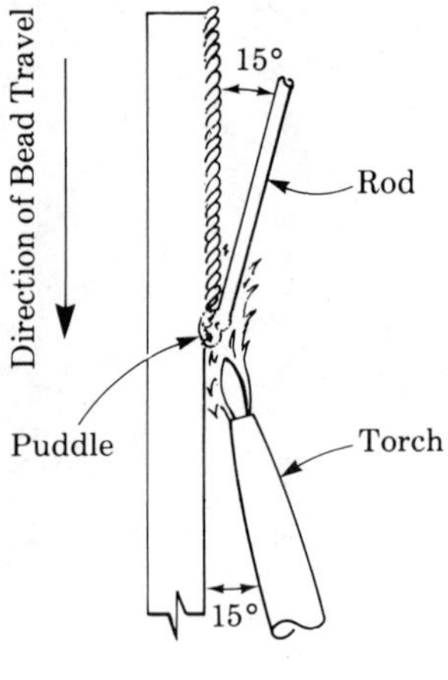

Figure 10-7.

OTHER VERTICAL JOINTS

The previous discussion related to the butt weld, but the fillet weld is also used on vertical welds. The major difference between the two is that it will require a larger puddle to fill the space between the right angle butt of the base metal.

The larger puddle will have more weight because of the extra filler metal required. You must control your torch manipulation more carefully to prevent drops from forming. You do this by keeping your puddle more mushy than you would for a flat weld bead. You will also have to weave your torch more from side to side to give proper bead coverage to each side of the two pieces of metal.

MULTIPLE PASSES ON VERTICAL WELDS

It is often necessary to make additional passes to give proper coverage. Such multiple passes are necessary if the metal thickness is too great to provide full bead height on the first pass. The number of passes required depends on the type of joint and thickness of the metal.

As explained earlier, a root pass is first laid to join the metals together. This initial pass must have full penetration to the depth of the groove between the metals, then a second pass is laid over the root pass. If the second pass fills the joint with a small build-up above the surface of the metal plate, you are finished.

If the joint requires more than these two passes, it is necessary to stagger the passes in a reverse pyramid form. Build on each side of the bead margin until the successive passes finally overlap with the correct bead height to fill the joint. This applies not only to flat position welding, but to vertical up and down welding as well.

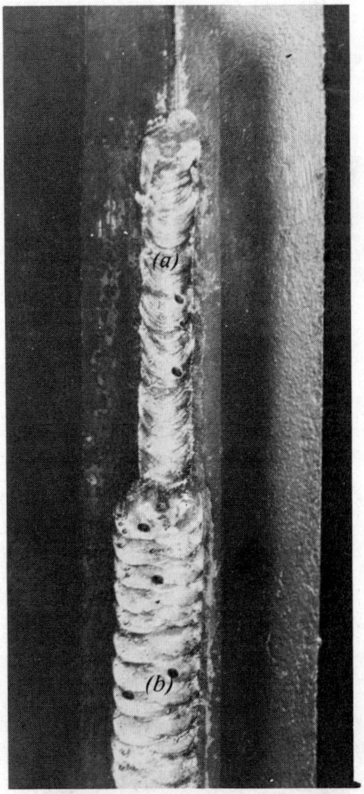

Figure 10-8. Multiple passes are made with a root pass (a), follwed by one or more successive passes (b) to build the bead to the proper height.

HORIZONTAL WELDING ON A VERTICAL SURFACE

The basic problem is to prevent the weld bead from sagging. To do this on a horizontal butt weld on a

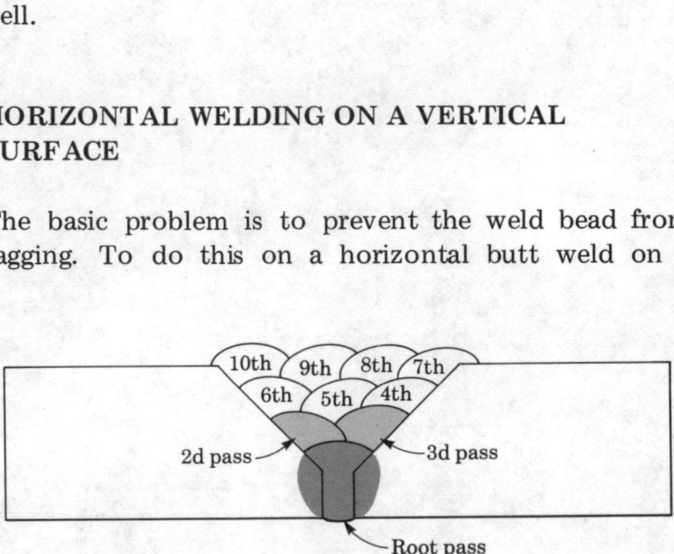

Figure 10-9. An exaggerated example of reverse pyramid multiple passes. Very seldom will more than three passes be required.

vertical surface, you must be careful of your torch angle. The flame must be directed upward at a 10° slant. At the same time you should apply the rod in such a way that it helps lift up the puddle.

The torch flame is directed toward the bottom of the metal joint. Heat rising from the flame and the heated bottom of the joint will heat the upper edge of the joint. You can judge when both become approximately the same temperature by their color. Ferrous metals turn cherry red just before reaching 1400°F, orange at slightly past 1700°, white hot at 2200°, and melting begins—depending upon the alloy—at 2600° to 2900°F.

Protect the weld puddle from overheating by holding the weld rod slightly in front of the torch flame. Deposit the filler rod right into the weld puddle and wipe up the weld puddle as you do so. This distributes the filler material into the joint.

Either the backhand or the forehand welding technique can be practiced with the horizontal weld bead. Pick the method that works best for you.

Figure 10-10. A lap weld is being welded in the horizontal position by using a backhand welding technique.

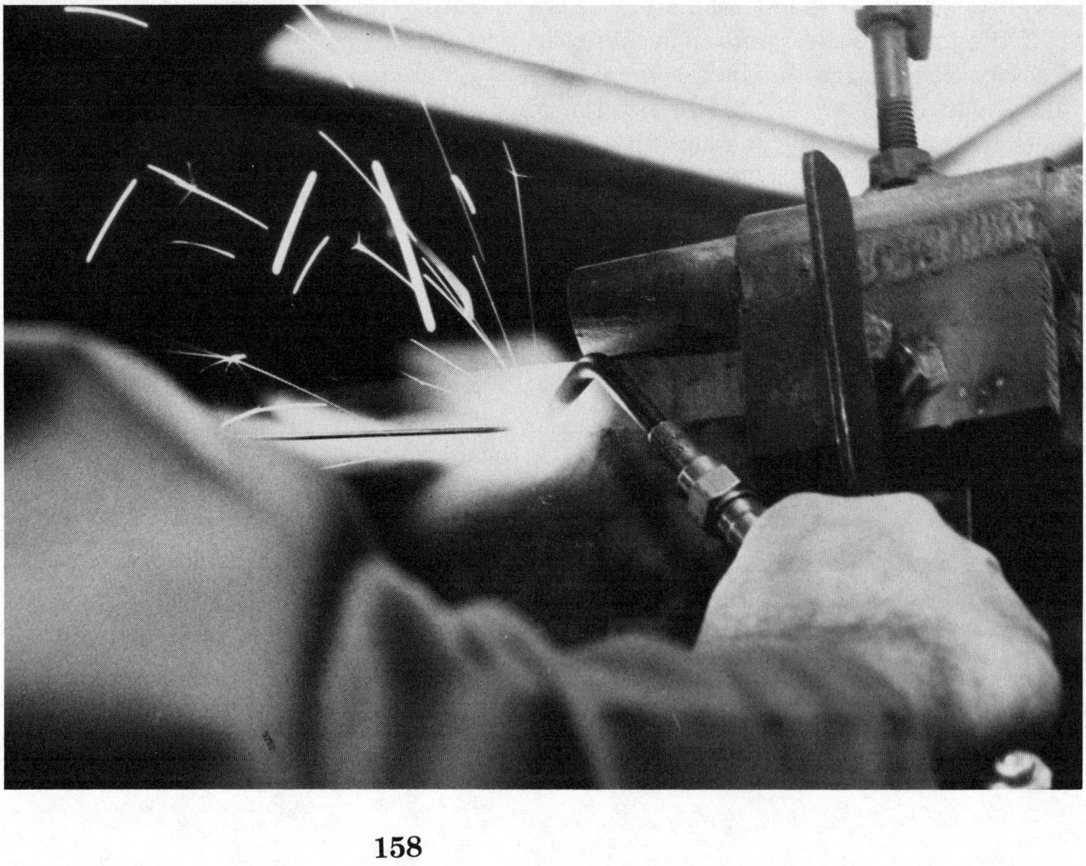

Lap and corner welds—horizontal on a vertical surface—are easier to make than the straight butt weld because the overlapping piece of metal provides a platform that helps support the bead. In fact, a horizontal lap weld on a vertical surface is no different from making a flat lap weld or T weld. Use the same technique.

Of course the metal must be vee grooved to secure good penetration if the metal is thicker than 1/8 in. Working within this groove will also give the base a support for the puddle that helps keep it from dripping and sagging; however, the slanting support of the bottom layer of the vee groove is not as good a support as the flat shelf provided by a lap weld. Again you depend upon the pressure of your torch flame gases and the uplift of your welding rod to keep the puddle in place.

OVERHEAD WELDING

We have saved the best position for the last—that is, if you can consider the most difficult position of all as the best. Still, overhead welding can be fun because it makes you really try harder. Because it is the most difficult of all welding positions, you should not attempt the overhead weld until you have thoroughly mastered the other positions. If you can make a good flat weld bead, vertical and horizontal weld on a vertical surface, you have gained the basic technical ability to make a good overhead weld. It will, however, require considerable practice.

Protect Yourself

The first and most important thing in making an overhead weld is to protect yourself. In working directly overhead, you are subject to sparks and dripping molten metal. You must wear protective clothing and gauntlet gloves and keep your pockets covered.

For your first overhead weld, position the metal about 12 in. above your head and stand off slightly to one side of the the weld joint. This will ensure, if something goes wrong, that the molten metal does not drip on your head. The most comfortable position is to stand with your shoulders parallel to the weld joint.

You will need to increase the torch angle to point almost directly into the joint. Heat all metal surfaces until the temperature increase is uniform in both sides.

When your puddle begins to form, start to melt your filler rod into the puddle.

Overcoming the Drip Problem

In the overhead weld we have the same problem that we have in the vertical weld. The puddle must be mushy and not too fluid. You must learn to flick your torch away at the proper moment to cool the puddle when it appears to overheat.

On an overhead weld you should not at first try to build up the bead to its full height. Work your torch and puddle about 1/4 in. ahead of your main puddle, then come back to the main puddle and add more filler rod to build the bead to the desired height. By working back and forth in this manner you will be able to build a full weld bead without actually forming molten droplets. You will find, however, that you must rely on the filler

Figure 10-11. A welder is positioned parallel to the weld bead. He is directing the torch flame into the weld seam to form the weld puddle. Control of the puddle is accomplished by both torch and rod movement.

rod instead of the torch flame to move the weld puddle around.

The torch flame can be used only to carry the minimum amount of heat to maintain the weld puddle just above the melting point of the metal. The process of capillary attraction will draw the molten filler metal into the joint and form your weld joint penetration.

After you have mastered the technique of the overhead butt weld, continue your practice with overhead lap welds and corner welds. Keep in mind the necessity for multiple passes when metal thickness requires it.

TEST YOUR JOINTS

After you have learned to make welds with good appearance and penetration, you should subject a few welds to a destructive test to check on your technique. Welded joints for sheet metal should be able to withstand a 90° bend. If the metal is low carbon steel of plate thickness, it may be bent 180° before it cracks or breaks.

Figure 10-12. The torch flame has preheated the metal to a bright red in this overhead weld. The sparks coming from the surface are a signal that the filler rod can be added to the weld puddle.

SUMMARY

1. Out-of-position welding means welding in any position other than the flat horizontal weld.

2. These include the vertical up and down welds, the horizontal weld on a vertical surface, and the overhead weld.

3. In all these out-of-position welds we are working against the force of gravity, which tries to make our weld bead drip down instead of staying in place.

4. Weld puddles are controlled by raising and lowering the torch and rod to allow the weld bead to partially solidify each time filler material is added.

5. The upward pressure of the torch flame also aids in preventing puddle drip.

6. Vertical down-welding makes use of a technique known as backhand welding. In backhand welding the torch flame is pointed in the opposite direction to the line of welding or torch travel.

7. The pressure of the torch gases plus the different heat concentration allow the weld puddle to be controlled more easily. This is due to the mushiness of the weld puddle.

8. Horizontal welding on a vertical base is most easily done with the torch flame slightly played against the bottom edge of the weld joint.

9. On these welds the filler rod is melted into the joint and wiped up to fill the joint as torch and rod movement are coordinated.

10. Multiple passes are used when the metal thickness prevents achieving the desired bead height on the first pass. A staggered reverse pyramid is used to build the passes.

11. For the overhead weld joint the torch angle is changed to point more directly into the weld joint. The welding rod is used to push the puddle around, and a back-and-forth torch movement is used to prevent a droplet from building up at the weld puddle.

12. Practice welds should be checked by a destructive test to ensure good penetration and fusion. A bend test in a vise is an easy test. If the weld joint can survive a 90° to 180° bend before breaking or cracking, the strength of the joint is adequate.

NEW WORDS AND TERMS

Backhand Welding A method where the torch is angled in the opposite direction of torch travel. Used in vertical welds and pipe welds.

Forehand Welding A method where the torch angle points in the direction of torch travel.

Horizontal Welding Welding a straight horizontal bead across a vertical surface.

Multiple Passes A method used to fill a vee groove with several layers of bead when one layer will not build the required height.

Overhead Weld The process of welding a joint positioned directly overhead.

Vertical Down Weld The process of welding a vertical joint by using a bead laid from top to bottom.

Vertical Up Weld The process of welding a vertical joint in which the weld bead is carried from bottom to top.

BRAZING AND BRAZE WELDING

In welding we melt two pieces of metal and let the molten material flow together, filling any remaining depression with filler metal of the same type as the base metal.

There are times when it is not desirable to melt the base metal, such as when raising the metal to the melting point would destroy special heat treatment in a portion of the material, making the weld area less strong than the rest of the material.

Another example of when melting the base material is undesirable is when we must join dissimilar metals. A metal gets its characteristics from the properties of its alloys. If we melt and fuse two dissimilar metals, within the fused area we will have an entirely different metal from either of the two we tried to join. This is because the fused area will have all the alloying materials of both metals. It might well be a very weak or porous weld.

In other cases it is not desirable to melt the base metal because of warpage or distortion.

UNMELTED JOINTS

When we do not want to melt the base metal we use brazing, braze welding, or soldering. All three of these processes, although differing somewhat from each other, are alike in that they join metal by melting only the filler rods. The base metal is heated, but the heat is well below the melting point, and sometimes below the critical range.

This lower range of heat prevents a change in the

grain pattern, and a corresponding change in the hardness of the metal lessens our problems in preventing distortion and permits us to join dissimilar metals.

DIFFERENCE BETWEEN BRAZING AND SOLDERING

The difference between brazing and soldering is that brazing is done at a temperature above 800°F but below the melting point of the base metal. Soldering is done at a temperature below 800°F. Also, a different filler metal is used in the two processes.

Brazing and braze welding use nonferrous filler metal such as brass, which is an alloy of copper and zinc, or bronze, which is an alloy of copper and tin. Soldering uses an alloy of lead and tin with a much lower melting point. (Soldering will be discussed in Chapter 12.)

BRAZING

If one could watch a brazing or braze welding operation, it would appear that it is done like oxyacetylene welding. This is true up to a point. The oxyacetylene torch and a filler rod is used but in a different manner (Figure 11-1).

Figure 11-1. The regular acetylene setup and torch are used in brazing.

First, the flame mixture varies according to what base metal you are attempting to braze. A slightly carburizing flame ensures less oxides forming during the heating process with nonferrous base metals, such as aluminum, brass, copper, or stainless.

With the ferrous base metals, the flame is returned to a neutral condition. If the base metal is very thick, a slightly oxidizing flame becomes helpful. This flame is used to heat the base metal along the joint. The temperature will be above 800°F; the exact heat depends upon the type of base metal being used.

The filler rod and torch, as in oxyacetylene welding, are held at a 45° angle to the work. The brazing filler rod is heated by the torch, but the rod must be kept outside the inner cone of the flame (Figure 11-2).

This type of indirect heat is also applied to the joint area. Gradually bring a section of the weld joint up to the proper heat as determined by the flux or melting characteristic of the filler rod.

Figure 11-2. In brazing, the rod and torch are held at a 45° angle as in oxyacetylene fusion welding. The torch heats the metal in advance of the rod. The brazing rod is melted by touching it to the red-hot metal heated by the torch. The molten filler is then drawn into the gap by capillary attraction.

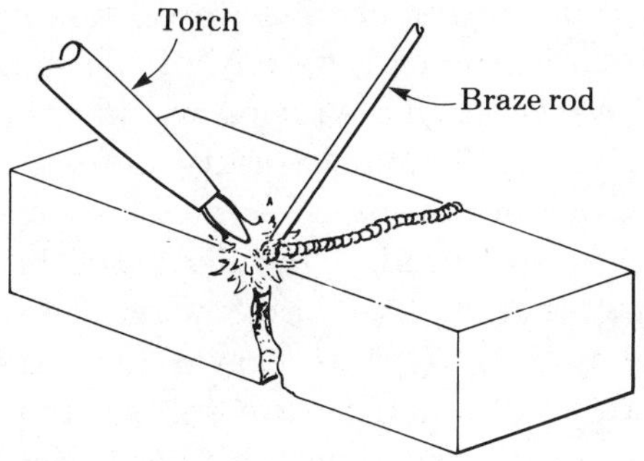

Figure 11-3. An illustration of the brazing principle. The torch heats the metal to a temperature below the melting point of the base metal. The brazing rod is melted by touching it to the hot metal. Capillary attraction pulls the flowing braze metal into the gap between the two pieces of metal.

The correct application of heat and filler rod becomes important as you begin the bead formation. If the metal is too hot, the filler rod will not adhere to the base metal easily. It will "ball up" like water on a hot griddle.

If the metal is too cold, your bead will become frigid or stiff. This will cause poor adhesion and little or no puddle fluidity.

In welding we melt the filler rod by dipping it in the puddle. In brazing there is no puddle since the base metal is not melted. The brazing filler rod is melted by touching it to the dull red steel heated by the torch and by the torch flame. The brazing rod has a melting point below the heat of the base metal.

The joints to be brazed are positioned close together, but not pressed tight. This leaves a slight gap between them. The melted filler rod is drawn into this tiny crack between the two pieces of base metal.

The filler is drawn into the crack by capillary attraction, and the filler and base metal grip each other by *molecular attraction*. This means that the molecules of the base metal lock with the molecules of the filler metal. This is the scientific explanation. For easier understanding, we might say that the filler and the base metals stick tightly to each other. In fact, they stick so tightly that a well-made braze joint may be as strong as a fusion weld. The braze joint is made as narrow as possible, but it must be wide enough for the molten filler to be drawn into the narrow gap.

BRAZE WELDING

Braze welding differs from brazing only in the type of joints used, the type of filler employed, and the application. In straight brazing we use the butt and lap joints. Braze welding can use any of the standard joints described for oxyacetylene fusion welding.

The American Welding Society defines braze welding as using a nonferrous filler metal to make a bead, fillet, plug, or slot bead. The filler metal has a melting point below that of the base metal. Unlike brazing, where the molten filler is drawn into the seam by capillary attraction (due to the surface tension of the hot metal), in braze welding the filler is deposited as in oxyacetylene fusion welding.

FILLER MATERIAL

Brass—a mixture of copper and zinc—is the most commonly used brazing filler material. Bronze—a mixture of copper and tin—is also used. However, brazing is not restricted to these two materials.

The development of unusual alloys for airplanes and rockets led to the need for brazing materials more suited to these so-called exotic metals. As a result, we now have rods made of nickel and chromium alloys, aluminum and silicon alloys, magnesium, copper, and others.

Silver brazing materials have been in use for centuries, having been developed by early jewelry makers. Today, silver brazing (known as silver soldering although it is not soldering) is an important industrial joining process and will be described in detail later.

The standard copper/zinc brazing rods are used because they have a high tensile (capable of being stretched) strength. In addition to the predominant copper/zinc content, these brazing rods may have small percentages of metals such as tin, iron, silicon, and sometimes manganese to improve their flow characteristics. The brazing material must flow smoothly and thinly between the gap of the base metals to achieve a good bond. This type of filler does not corrode.

FLUX

Before beginning a detailed explanation of brazing procedure, there is one other important point to lcarn. This is *flux*. Flux is defined as "a substance that helps metals fuse together." The American Welding Society's definition is more complete: "Flux is a material used to prevent, dissolve or facilitate removal of oxides and other undesirable substances."

To put it simply, flux is used during the brazing process to clean the metal and keep it clean.

Applying Flux

Flux comes in powder form or as tiny crystals. The welder can heat the bottom 2 or 3 in. of the brazing rod with his torch and then dip the rod into the flux can. The flux will stick to the hot metal. Care must be taken in heating the rod, however, to prevent melting it. The flux sticking to the rod is later melted into the filler as the filler rod is melted on the hot base metal.

Another way to apply flux is to heat the base metal along the intended joint. The flux is then sprinkled on the hot metal. It will stick the same way it does to the hot brazing rod. Then the braze or braze welding is done in the usual manner. The heat of the oxyacetylene torch melts the flux into the filler.

The dip method is most often used with brazing because it is not as messy. When the sprinkle method is used, the torch melts the flux all over the base metal.

Automatic Flux Application

The dip and sprinkle methods of applying flux are handy for small-quantity brazing and braze welding.

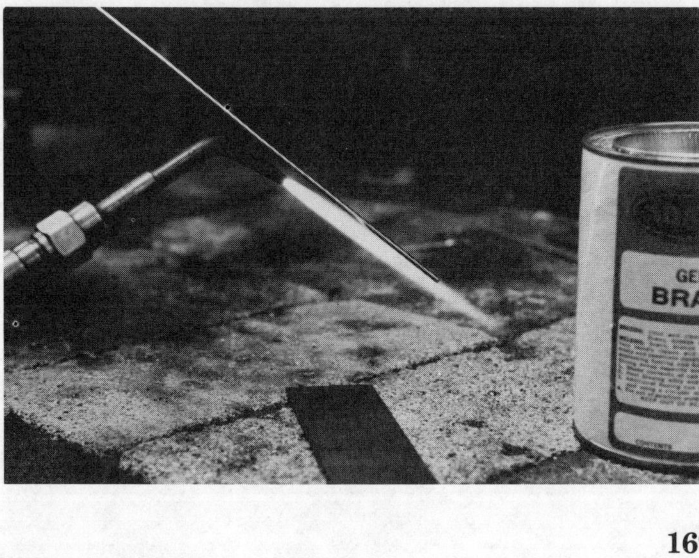

Figure 11-4. The brazing rod is heated in the torch flame to make it hot enough to pick up flux.

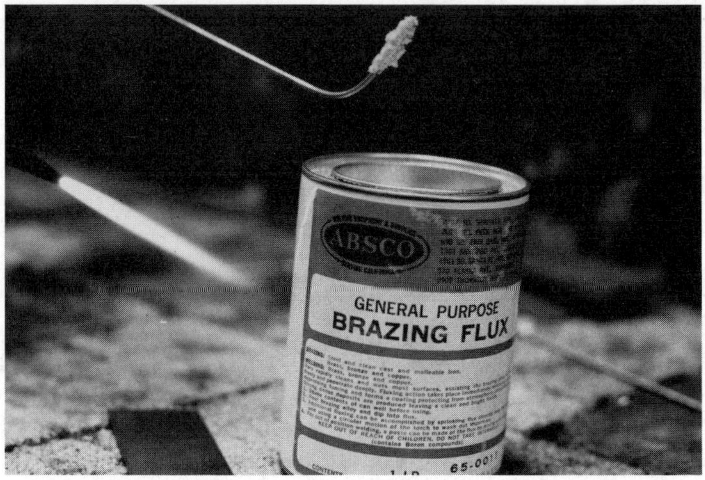

Figure 11-5. Flux sticks readily to the hot brazing rod.

Shops handling large amounts of such work often use automatic flux application. In these methods the flux is fed into the fuel gas (acetylene or other fuel gases) and comes out through the torch tip with the flame. One such method dissolves the flux in alcohol. The mixture is then fed into the flame through the torch.

Brazing rods can be purchased with flux coatings. The flux covers the full length of the rod so that dipping is not necessary. This is most useful and the easiest method since it does not require constantly interrupting the flow of work to dip the rod.

Some welders coat larger portions of their rods (Figure 11-6) to avoid constant dipping. This can be done by pouring some flux in a trough (a piece of angle iron

Figure 11-6. A longer length of rod can be coated by spreading the flux in a receptacle like this piece of angle iron.

is good), heating the rod for most of its length, and laying in the flux to coat almost the full rod.

What Flux Does in Brazing

Earlier we defined flux as material used to clean metal and keep it clean. The question often arises with beginning welders as to why the metal cannot be cleaned with grinder or wire brush as we prepare iron and steel for fusion welding. One reason is that the metal must be cleaner for brazing than for fusion welding. The heat used in brazing is not as hot and there is less chance to burn away impurities and oxides. These can interfere with a good braze. The metal must be chemically clean before the brazing filler will stick to the base metal.

You have noticed how water collects in balls upon the surface of a waxed car instead of spreading in a thin film. The same thing can happen with brazing materials when used on improperly prepared metals. This prevents full or even molecular adherence. Flux removes oxides that cause this trouble and permits a good bond in the joint.

Cleaning During Brazing

It is not sufficient to clean metal. It must be kept clean all during the brazing process. This is another job flux does for us. It cleans away oxides that form during the brazing process.

Oxygen in the oxyacetylene flame and oxygen in the air surrounding the joint both combine with elements in the molten brazing material when it is melted by the oxyacetylene flame. Flux removes the resulting oxidation. Oxidation, you will recall, is the product of oxygen combining with a metal.

Composition of Flux

The composition of flux depends upon the type of metal being brazed and the kind of brazing filler used. Borax is the traditional flux used with copper alloy fillers. Since manufacturers have their own formulas for

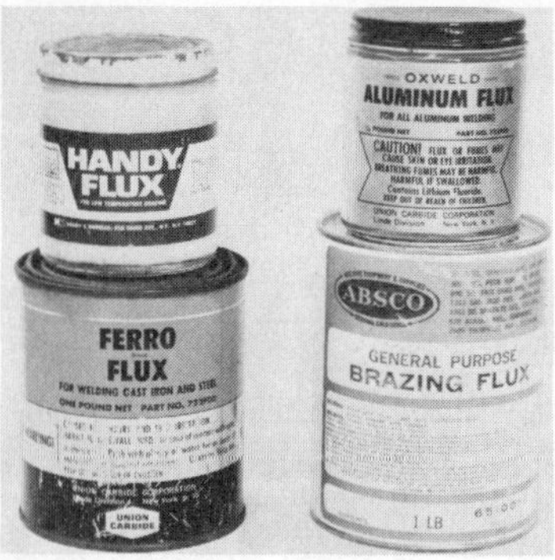

Figure 11-7. Fluxes come in different compositions to meet requirements of different metal joining.

fluxes, these may contain, in addition to borax, small amounts of iodine, bromine, chlorine, and other substances added to aid cleaning and flow characteristics of brazing fillers.

Stainless steels, aluminum, and certain alloys require different fluxes. Alkaline bifluoride is used with these.

Safety Precautions

Some fluxes, notably sodium cyanide (used in one form of silver soldering), give off poisonous fumes when heated in the brazing process. In all cases *adequate* ventilation must be provided and care taken not to inhale any of the fumes.

Zinc and cadmium, used in some metal alloys and filler rods, also give off toxic fumes when vaporized in brazing. Cadmium is especially deadly.

PREPARATION OF METALS FOR BRAZING

Brazing is the thin flow of a binding filler between the pieces to be joined, so only the butt and lap weld are used in this process.

Preparation of the metal is also different from fusion welding. The two pieces to be joined are separated by a gap of 0.004 to 0.010 in. This gap provides the thin opening for the brazing material to flow into and stick the two pieces together.

The base metal must be held rigid by clamps during the brazing process until the metal has cooled.

Figure 11-8 shows how butt and lap welds are made in brazing. In (*a*) the two pieces of metal for brazing are clamped to a support with C clamps. The gap between them (exaggerated in the drawing for clarity) is set at 0.004 to 0.010 in. In (*b*) the gap has been filled with brazing filler and left—*still clamped*—to cool.

Figure 11-8(*c*) and (*d*) show how a butt weld is made. The two pieces are overlapped and clamped to a support as shown in (*c*). A block of metal is placed under the top lap to keep it raised to the proper height to provide the necessary clearance between the two pieces. The pieces are then clamped in place to prevent them moving during the brazing process; (*d*) shows how the

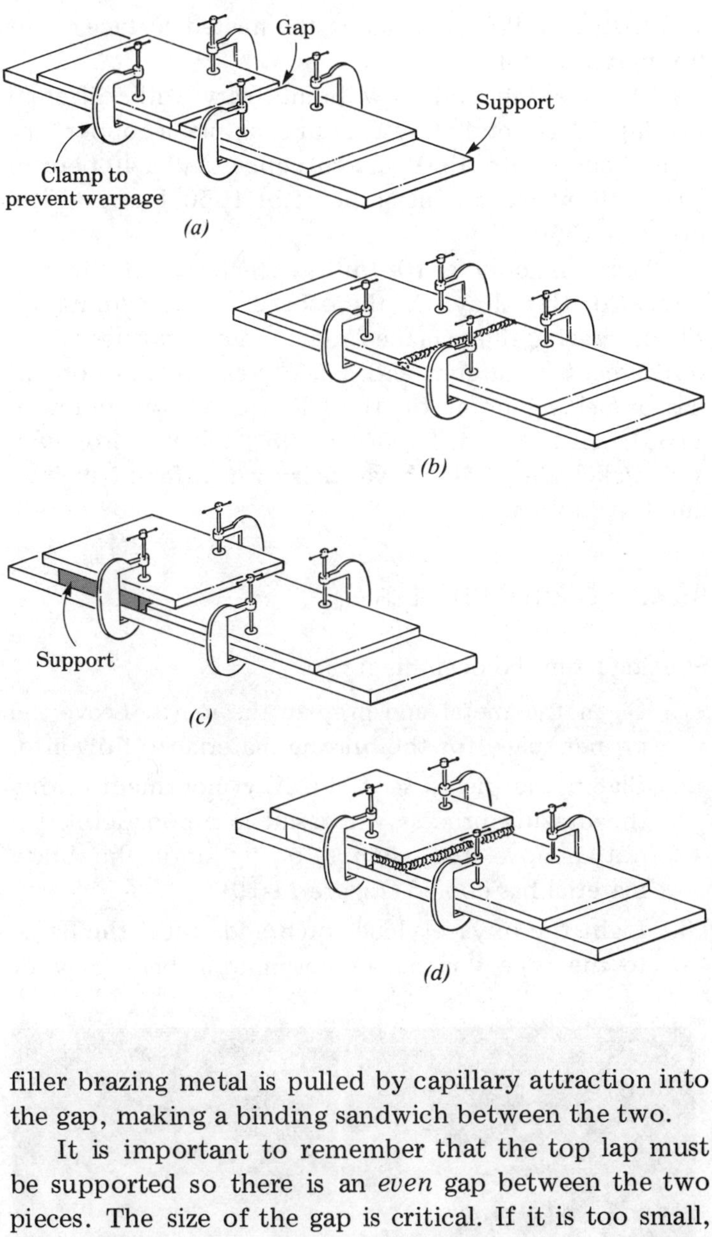

Figure 11-8. The lap (a), (b) and butt welds (c), (d) used in brazing.

filler brazing metal is pulled by capillary attraction into the gap, making a binding sandwich between the two.

It is important to remember that the top lap must be supported so there is an *even* gap between the two pieces. The size of the gap is critical. If it is too small, there will not be complete penetration. If it is too large, the braze joint will be weakened.

Flow Point

It is important to distinguish between the melting point of brazing filler and its *flow point*. Metals melt before they flow. Therefore, the temperature of the filler must be raised above the melting point to the flow point. It will then run like water along the heated surface of the hot parent metal.

The melting and flow points vary with different brazing filler metal alloys. As an example, a copper/zinc (brass) alloy with a 60 percent copper and a 40 percent zinc content has a melting point of 1650°F and a flow point of 1660°F.

There is only a 10° difference between the two points for this alloy. On the other hand, an aluminum/silicon brazing filler (made for aluminum brazing) with a 95 percent aluminum and a 5 percent silicon content has a melting point of 1070°F and a flow point of 1165°. This is a difference of 95°. Silver, chromium, and nickel alloys also have their own different melting and flow points.

BRAZING PROCEDURES

Starting from the beginning:

1. Clean the metal and prepare the joints. Leave the proper space for the brazing material to flow into.
2. Clamp the pieces securely. Any movement during the brazing process will result in a poor joint. Do not remove the clamps or jig until the braze material has cooled (Figure 11-10).
3. Light the oxyacetylene torch and adjust the flame to the type you have determined is best for your

Figure 11-9. Brazing is done by touching the brazing rod to the heated metal.

base materials. Nonferrous metal does better with a carburizing flame. Ferrous materials use either a neutral or slightly oxidizing flame.

4. For safety, make sure that you have adequate ventilation.
5. Gently heat the base metal along the area to be brazed.
6. Heat the brazing rod by holding it inside the flame envelope, but *not* inside the inner cone, or it will heat too fast and cause the rod to melt.
7. After the rod is heated, dip the heated rod into the powdered flux. A longer section can be coated by rolling the rod in a trough, such as that made by a piece of angle iron.
8. The metal to be brazed must be heated just slightly above the melting point of the brass filler rod. A dull red glow in ferrous metal is a signal for the approximate degree of heat for the flow point of the filler rod.
9. Now that the base metal is glowing a dull red, bring the torch and the filler rod in quickly and "bite-off" a piece of the filler rod with the torch flame. Continue heating around the joint area and adding the filler rod.
10. The rod and torch can be manipulated with a coordinated movement in a circle or forward and backward.

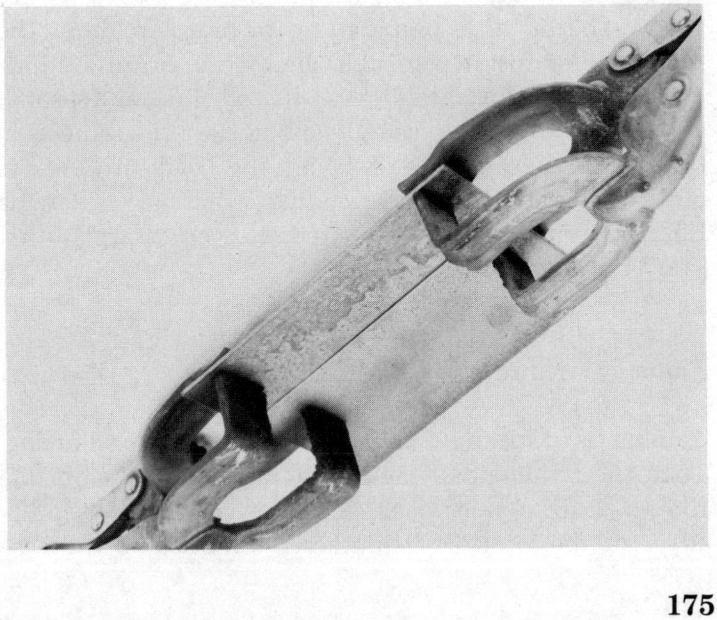

Figure 11-10. Clamping the metal before beginning the bead will help eliminate warpage.

11. If the metal is not heated enough, the filler rod will pile up. It will not flow since the heat of the metal is below its flow point.
12. If the metal is too hot, the filler rod will melt and explode like water beads on a hot griddle. The filler rod will spread out very thin. A white coating will form on the bead margins. This white is the zinc alloy in the rod.
13. Between these two extremes the filler rod will melt and flow evenly. There will be a minimum of zinc coating and no splatter.
14. An important point to remember is that the thinner the layer of brazing material between the two pieces of parent metal, the stronger the braze will be.
15. The brazing filler must penetrate completely through the seam.
16. The final step is to cool the joint and inspect the joint for complete penetration and defects.

BRAZE WELDING

In brazing we flow a thin layer of filler between the parts to be joined. But in braze *welding* we lay down a bead of filler metal just as we do in fusion welding (Figure 11-11). The only difference is that we do not melt the base metal.

As in fusion welding, it will be necessary to vee groove the metal in preparation for braze welding. The full range of fusion joints can be used in braze welding.

The need for cleanliness, use of flux, and proper adjustment of the oxyacetylene flame are the same as in brazing. The technique of laying the bead, however, is quite different. In braze welding, where considerable filler material must be deposited, it is necessary to first "tin" the edges of the joint.

Tinning

Tinning is done by heating the metal as for brazing. Then the red-hot base metal is touched with the brazing rod to flow a thin layer of brazing material over the edges of the joint. The thin layer will stick to the base metal where a heavy flow—needed to fill a vee groove—will adhere poorly.

Tinning does not necessarily have to be done on the entire groove at one time. A skillful welder can tin just ahead of his full weld bead.

As in brazing, the filler rod is not melted by the flame. It is heated in the flame envelope, but kept from the inner cone. The heated and fluxed rod is then melted by touching it to the flame-heated base metal (Figure 11-12).

Laying a Braze Bead

In laying a bead with brazing filler, the torch moves in a slightly circular motion ahead of the rod to heat the

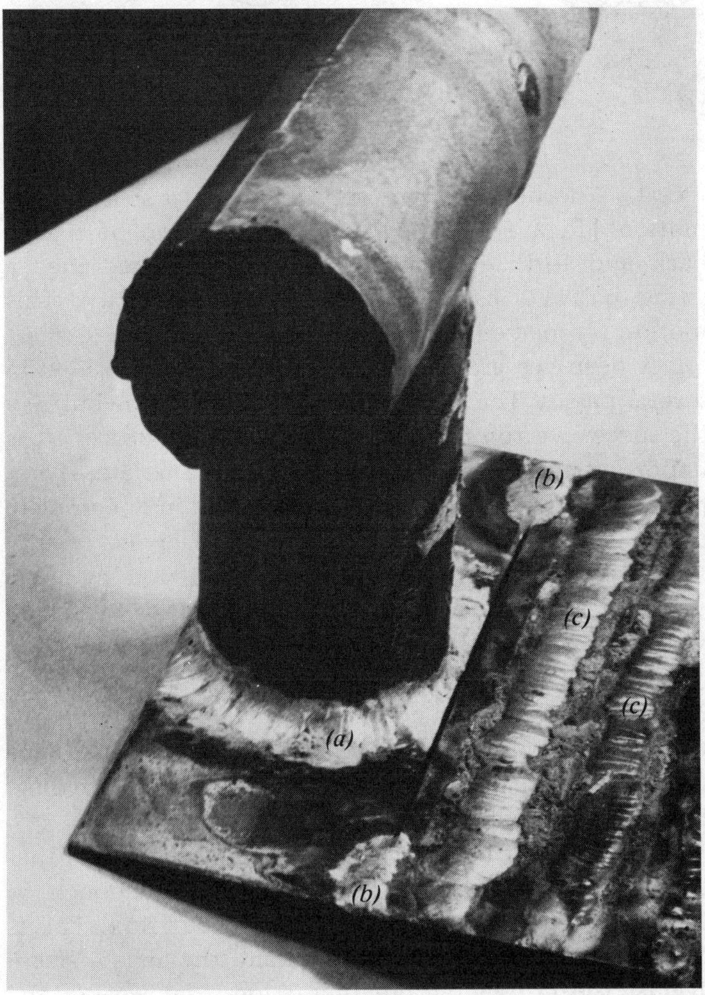

Figure 11-11. In braze welding, a bead is laid down just as in fusion welding. This illustration shows a fillet bead (a), two tack welds (b), and two flat horizontal beads (c).

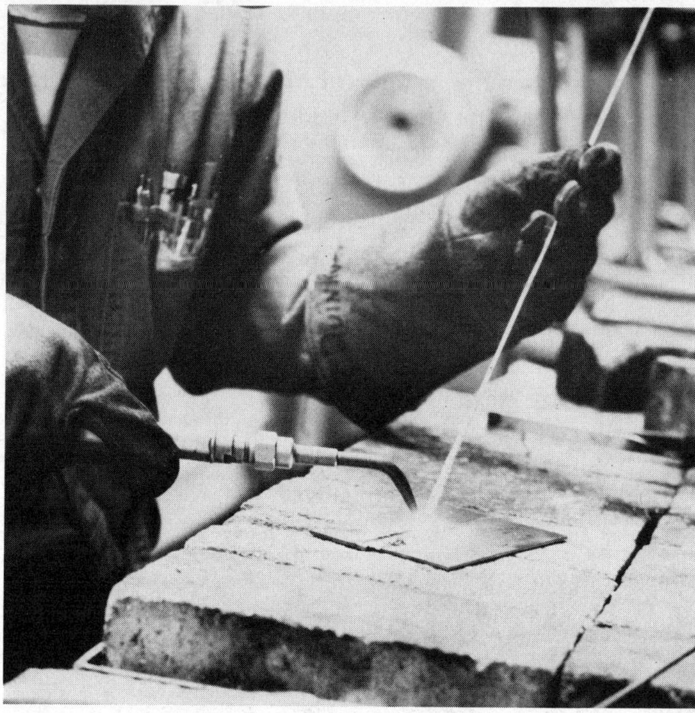

Figure 11-12. The flux-coated filler rod is melted in braze welding by touching it to the heated metal in the same manner as for brazing.

metal. Another popular movement is a semicircular motion [Figure 11-13(*b*)], which is a kind of zig-zag back and forth, or a semicircular advance and then a swing halfway back up the previous swing (*c*). This motion is repeated for the duration of the bead.

A deep vee groove may have to be braze welded in several passes. The first pass would be to tin the butts so the successive rows of beads would stick properly. It is important to ensure that this tinning is complete. If any parts are left untinned, there is danger of an incomplete braze. It will be weak and possibly fail in service.

After the vee groove has been tinned, you lay down your first bead. This is done exactly as in fusion welding, except that you do not melt the base metal.

The tinning was handled in a thin coat like brazing. Now we work with puddles. Melt sufficient rod to make the initial puddle. From that point melt the braze rod by dipping it in the puddle, as in fusion welding. Proceed by building overlapping puddles.

Continue to use flux. Many beginners assume that the tinning, having been melted, is clean enough to proceed with the braze weld pass without flux. This is not true. Oxygen from the torch and the air are active against the hot metal, creating oxides that weaken the

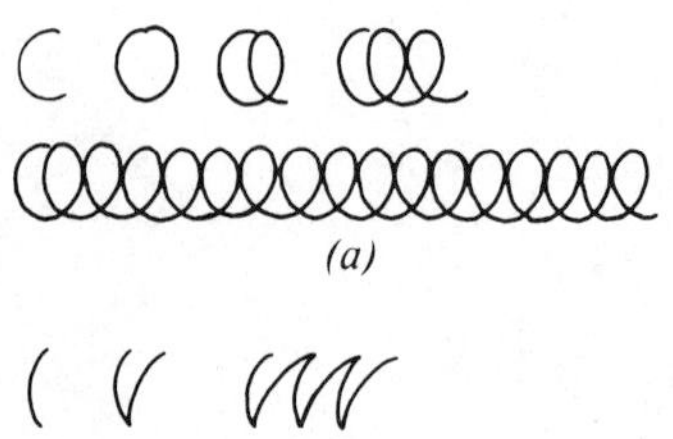

Figure 11-13. Torch movements in laying a braze bead.

braze weld bead. Flux is necessary to remove these oxides.

The bead is carried by using one of the forward motions listed above. If more than one pass is necessary to fill a deep groove, each successive pass must fuse into the previous one. The filler does not fuse into the base metal, but it must fuse into the previously laid filler bead.

SUMMARY

1. Brazing and braze welding join metals with a nonferrous filler with a lower melting point than the base metal.

2. Brazing "sticks" the base metal pieces together with a thin film (from 0.004 to 0.010) that is drawn by capillary attraction between the close lap or butt joints.

3. Braze welding joins thicker metals that must be vee grooved to provide adequate penetration.

4. Vee-grooved metals must be "tinned" before braze weld beads are deposited.

5. Brazing and braze welding are used to join dissimilar metals. They are also used for joining where excess heat will produce undesirable changes.

6. Filler rods for brazing processes are alloys. The most common alloy is brass, made of copper and zinc. Bronze—copper and tin—is also used. Other alloys, used for aluminum brazing, are made of aluminum and silicon.

7. Flux must be used to chemically clean metals for brazing. It also removes oxides created during the brazing process.

8. Borax is a common base for flux. Flux can be deposited on the filler rod by heating and dipping the rod in a can of the material, sprinkled on the hot base metal, or introduced into the flame. Flux-coated rods are also available.

9. Some fluxes give off poisonous fumes when heated. It is imperative that adequate ventilation be provided when brazing.

10. Butt and lap welds are used for brazing. Braze welding can be done with any of the standard fusion welding joints.

NEW WORDS AND TERMS

Braze To join similar or dissimilar metals by using thin films of nonferrous metal with a melting point above 8000°F.

Braze Welding The use of nonferrous filler material with a melting point above 8000°F to make a groove, fillet, or flat bead on an unmelted base metal.

Flux Chemical preparation, often of a borax base, used to clean metal and prevent oxidation during brazing processes.

Molecular Attraction The action of metals that causes the brazing material and the base metal to grip each other through the interlock of their molecules.

Molecules The smallest unit of an element before it breaks into atoms.

Solder To join unmelted base metal with a filler material that melts below 800°F.

Tinning The process of coating a braze joint with a thin flow of metal filler to act as a binder for succeeding heavy beads.

SOLDERING

Soldering with the oxyacetylene torch requires a cooler carburizing flame because the filler materials are lead and tin and the flow temperature is below 800° F.

The exception to this is silver soldering, which is really a form of brazing since it uses a brazing temperature above 800° F.

SILVER SOLDERING

Silver soldering was invented by jewelers, but in recent years it has found increasing applications in industry. Silver is very expensive, and general silver welding is rarely done. Silver soldering requires less silver filler.

Silver soldering is exactly like regular brazing; that is, the pieces of base metal are fitted together with a gap of from 0.004 to 0.010. The metal is heated above the flow point of the silver solder but below the melting point of the base metal. The silver solder is melted by touching it against the hot metal. Capillary attraction pulls the silver solder into the joint just as it does in brazing.

Composition of Silver Solder

The silver filler metal is an alloy of silver with other metals, usually phosphorous, cadmium, zinc, or copper. Gold is sometimes used, but it is so expensive that it is used industrially only for soldering certain precision instruments. The amount of actual silver in these alloys varies from as little as 9 percent to as much as 80 percent.

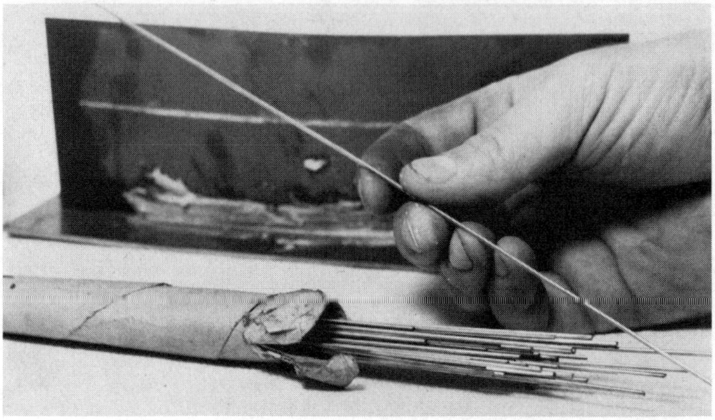

Figure 12-1. Silver solder is an alloy of silver with other materials. Here a silver solder rod is being picked from a package of Sil-Fos silver solder.

Different alloys have different melting and flowing points which must be considered when silver soldering. In some cases the different metals in the alloys will not melt at the same temperatures and can cause separation. The lower melting point metals in the alloy will braze into the joint before the other substances can reach their melting point. This can be avoided by heating the metal as quickly as possible.

A Safety Precaution

Cadmium and zinc, used in silver solder alloys to improve flow qualities, produce toxic fumes, especially if they become overheated. The welding booth or area must be well ventilated when using solders containing either of these elements.

Flux

Joints for silver soldering must be as thoroughly cleaned as those used for brazing. Flux is required as a chemical cleaner on the metal surfaces. The metal is first mechanically cleaned with emery cloth and then a thin film of paste flux is brushed over the area to be soldered.

The flux not only serves as a cleaner but is a temperature indicator as well. By watching the flux change color and consistency, you can determine when the metal temperature is about ready to melt the silver solder.

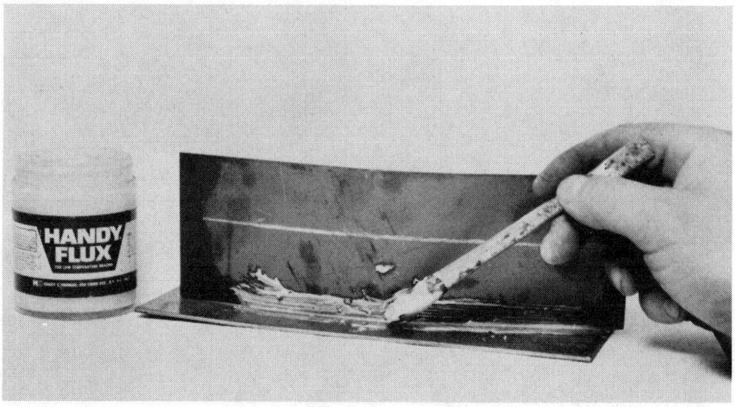

Figure 12-2. Handy Flux is a brand of silver solder flux. It is being applied here with a brush applicator to a T joint. This flux is also used as a temperature indicator and as a cleaner.

Heat Range

The melting point for silver solders is in a range of about 1050° to 1100°F. It is important not to heat the metal beyond this range because of the toxic fumes that form from the cadmium and zinc in the flux. The flow characteristics of the filler metals are also affected by the higher temperature range. The filler spreads too thin and material is wasted.

If the temperature is below the flow point, the filler metal will become mushy and will not flow easily. Temperature-indicator flux then becomes a useful way to tell when to add filler metal.

Action of the Flux

The flux turns milky and chalky looking at about 600°F. There are some bubbles that appear at this time, and at the correct temperature that the flux is designed to indicate, the chalky white color turns to a clear liquid, like a film of water. When you see this change, it will be your signal to add the solder.

SILVER SOLDER PROCEDURE

Silver soldering uses a reducing or carburizing flame, which has more acetylene and less oxygen than the other basic flames. Otherwise, the silver soldering procedure is the same as that given for brazing:

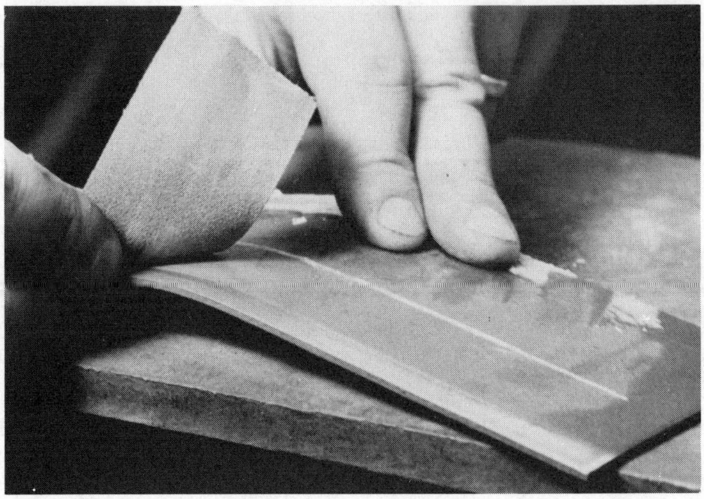

Figure 12-3. Emery cloth is used as an abrasive to clean the edge of this metal prior to the metal being silver soldered.

1. Clean the metal thoroughly—emery cloth may be used (Figure 12-3).
2. Fit the pieces with a narrow gap and clamp securely. The thinner the gap the stronger the joint.
3. Select a flux suitable to the metal being joined.
4. Adjust the oxyacetylene torch to a reducing flame.
5. Heat the metal to the flow point of the solder alloy.
6. Apply solder by touching it to the hot metal.
7. Leave the pieces clamped until the solder cools.

Importance of Cleaning

Fluxes used in silver soldering have a corrosive effect; therefore the bead must be thoroughly cleaned by washing in clear water. After a thorough cleaning, inspect the joint for penetration and proper bead.

REGULAR SOLDERING

True soldering is done at a temperature below 800° F. It is the only way that metals with low melting points can be joined.

The oldest and best known soldering technique uses a "soldering iron." This iron, with its copper tip, is heated and used to transfer heat to melt the solder. The soldering iron may be heated in an open flame, by a

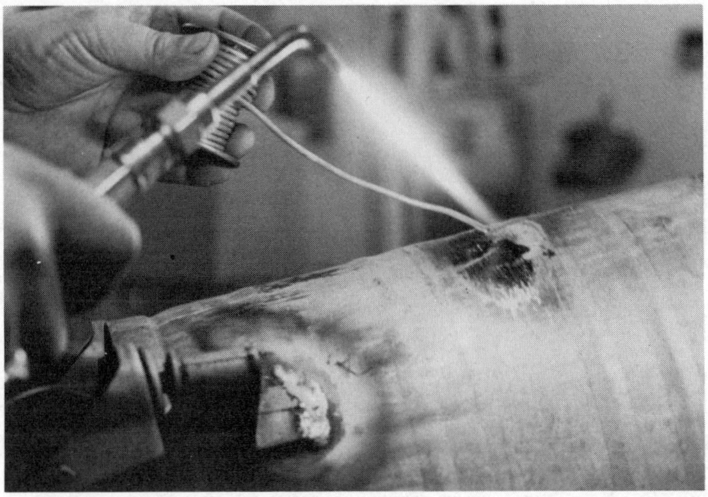

Figure 12-4. Solder is melted by touching it to the metal heated by the torch.

blow torch, or by electricity. Soldering can also be done by using a carbon arc, by dipping the parts in a solder bath, or by using an oxyacetylene torch. Since we are concerned here with the use of oxyacetylene, the torch method of soldering will be the only one discussed in detail.

How Solder Works

Solder works on the same principle as brazing. The base metal is heated above the flow point temperature of the solder and the filler is melted by touching it to the hot metal. It binds the same way that brazing filler does (Figure 12-4).

Solder is basically lead and tin alloyed but may contain traces of other metals to adapt it for different uses. Solder is an excellent electrical conductor and is used for joining electrical wires. It is not, however, as strong as a braze or welded joint but is generally weaker than the metal it joins. Also, a soldered joint cannot be used for any connection that will be heated above the flow temperature of the solder.

The flow temperature varies. A solder made with 25 percent lead, 25 percent tin, and 50 percent bismuth has a flow temperature of only 266°F. High-temperature solder is composed of 5 percent tin and 95 percent lead. It has a flow temperature of 595°F. General-purpose lead/tin solders are compounded for flow temperatures ranging from 361° to 500°F.

Figure 12-5. In some solders the flux is contained in the core of the solder. This is an example of acid-core solder.

Solder Flux

Solder also requires flux, and there are basically two types: acid and rosin. Acid flux, the most common, has a zinc chloride and ammonium chloride base and may be used in liquid or paste form. Under heat the flux forms hydrochloric acid that dissolves oxides on the base metal, permitting the solder to stick.

Solder flux is corrosive and must be cleaned from the solder bead. Brushing is not enough. The soldered joint should be washed and brushed with a wire brush.

Electrical and electronic connections require a noncorrosive flux. Rosin, a resin by-product of pine trees, is used for these solder joints. Although rosin is considered a noncorrosive material, some rosin fluxes have a dilute hydrochloride addition to help the rosin spread more quickly. This gives a slightly corrosive effect. Joints soldered with such a flux must be thoroughly cleaned.

Ammonium chloride fluxes are used for soldering copper and iron. Zinc chloride is used for sheet metals, copper, and brass.

TORCH SOLDERING

1. Clean the metal surfaces thoroughly. Steel wool is a good cleaning agent.

Figure 12-6. The metal surface is being cleaned with steel wool to prepare it for a soldered joint. The steel wool removes all dirt and oxidation from the metal.

2. Use proper flux and adjust torch for carburizing flame.
3. Heat the metal to be soldered. Touch solder to the hot metal as in the brazing process.
4. Avoid touching the solder with the torch.
5. The applied solder bead can be smoothed by sweeping with the torch.

Unlike a weld bead, if too much solder has been deposited, it can be wiped off while the solder is still molten. Wear gloves and use a thick cloth folded several times. Remember that the heat of melted solder is two or three times that of boiling water.

You do not need much heat for solder, so be careful not to overheat either the metal or the solder. Keep the torch farther from the base metal than you did in fusion welding. Keep the torch moving all the time. This will help avoid overheating the base metal. A damp rag can be used to cool the solder quickly when the joint is completed. This will set the solder joint quickly and firmly.

SWEATING

Sweating is a special soldering technique. It means that solder is applied to the edges of two pieces of metal and

Figure 12-7. A damp cloth can be used to quench the heat from the soldered joint and set it up quickly.

hardened. Then this tinned area is overlapped and the solder between the pieces is melted by applying heat through the base metal. Sweating is often used to solder in positions where it is difficult to apply the solder in the regular manner. The procedures for sweating are:

1. Tin the portions of the metal that will overlap. Tinning means to spread a thin layer of solder over the metal [Figure 12-8(*a*)].
2. The area containing the hardened solder is then overlapped [Figure 12-8(*b*)] and the tinned layers remelted by heating the outside of the top layer of base metal. This fuses the two layers of solder together.

It takes some experience to determine how much heat should be applied to penetrate the base metal and fuse the tinned layers. The major problem is to avoid overheating or melting the base metal.

Sweated Joint

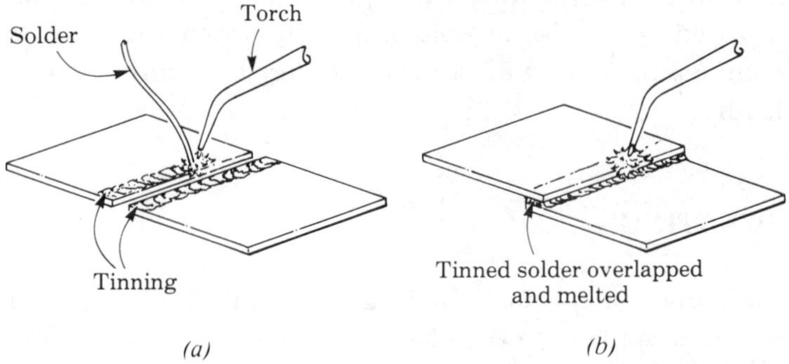

Figure 12-8.

Figure 12-9. These copper pipes can be joined to fabricate water systems in plumbing. This particular joint is used to operate a home water conditioner.

SOLDERED COPPER PIPES

Soldered copper pipes are often used in various types of plumbing. The soldered joint gives a leakproof joint that is especially desirable in air conditioning and refrigeration.

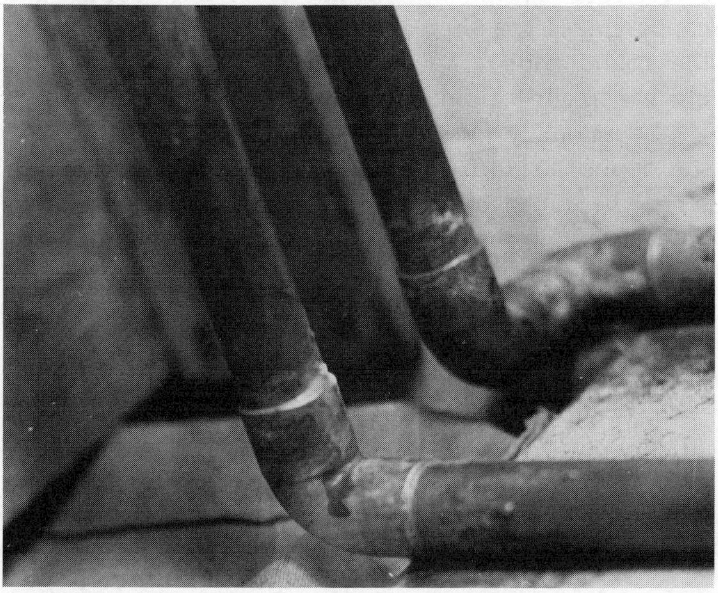

Figure 12-10. These copper pipes have been soldered together by use of an air-acetylene torch. This method is very quick and commonly used in plumbing repair work.

Soldered Pipe Joint

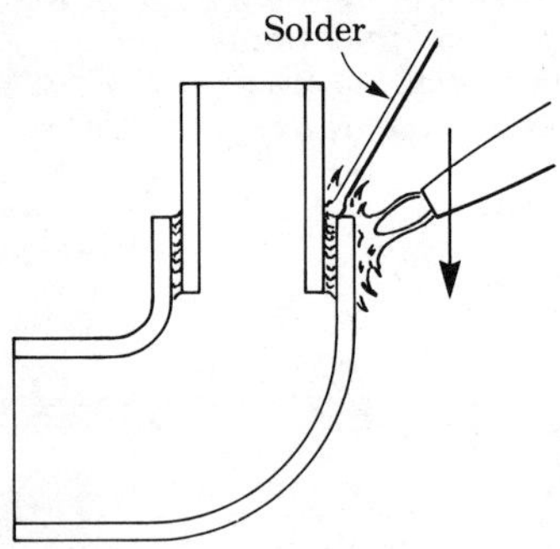

Figure 12-11.

The copper pipe or tubing to be soldered is made to fit inside another pipe with a very small clearance between the two. The clearance is about equal to a brazing gap. The joints are thoroughly cleaned with emery cloth or steel wool, and the joint is coated with rosin flux before the two pieces are fitted, one inside the other. Then the joint is heated and solder is applied (Figure 12-11).

The solder is introduced into the gap between the two pieces of copper pipe in the same manner as brazing; that is, the joint is heated and the solder touched to the outside edge. Capillary attraction then pulls the flowing solder into the heated gap.

ACHIEVING A STRONG JOINT

To achieve a strong copper pipe joint, the surfaces of the two metals must be cleaned and fluxed. The easiest way to introduce the flux is to brush it on the outside of the inner pipe. Flux is not put on the inside surface of the outer pipe because we must have only a small gap—0.004 to 0.010--between the two pipes. If both mating surfaces are fluxed, it would be too much for so small a gap.

The entire gap must be heated evenly to allow the solder to be drawn into the gap uniformly. Set the two copper pipes together and begin heating all surfaces of the round pipe evenly. Move the torch flame in a manner to allow full heat to transfer to all surfaces.

Occasionally touch the solder to the copper pipe on the opposite side from that which the torch is heating. When the solder melts easily, remove the flame from the pipe and touch the solder quickly around the entire pipe joint. The solder should melt evenly.

Do not add too much solder; let a small amount be drawn into the joint. Apply the flame again just briefly to keep the heat uniform. If you reheat the joint excessively, the solder will flow out of the joint. Apply a damp cloth to the joint to quickly set the solder. This sudden cooling will set the copper pipe firmly together. A rag should be used to wipe away any excess rosin that melted and flowed around the joint.

SOLDERING STAINLESS STEEL

Stainless steel can be soldered if a low-strength joint is satisfactory. The composition of stainless steel varies according to the amount of chromium and nickel alloyed with the steel, and special solders are required.

Chromium resists ordinary fluxes. A recommended cleaner for stainless steel consists of a liquid solution of hydrochloric acid and water, mixed half-and-half. A beginner should not mix this solution without experienced supervision. There can be a violent reaction when water is added to the acid. Always add the acid to the water slowly. Wear rubber gloves and protection for your face and body.

After the stainless steel has been treated with this corrosive cleaner for 3 to 4 min., rinse it with water. Then use a zinc chloride flux for the actual soldering. Solders can be bought with an alloy balance to match the exact color of stainless steel.

ALUMINUM SOLDERING

Aluminum is difficult to solder. Pure aluminum can be soldered, but certain kinds of clad, heat-treated, or aluminum alloys cannot.

A beginner should not attempt to solder aluminum. Instead, this metal should be welded. If aluminum must be soldered, then special solders are required. Lead, an important ingredient of ordinary solder, does not adhere well to aluminum. Special aluminum solders and fluxes are available from welding supply houses.

SAFETY PRECAUTIONS

Certain fluxes and solders give off toxic fumes. Good ventilation is essential to all soldering operations. Flux is corrosive. Any flux spilled on the hands must be flushed off with water as quickly as possible.

Soldering is a low-heat process that is ideal for use in the home garage. Unfortunately, it is a source of fires due to careless handling of torch and soldering iron. The area where soldering is done must be kept clean. Grease, oil spills, and trash ignite easily. Soldering should never be done in a dirty or potentially explosive atmosphere.

Under no circumstances should any pipe or container be soldered if the articles have been used previously to hold flammable substances. Residues and fumes are nearly always present. The heat of the torch or soldering iron may cause an explosion. Such items can be welded or soldered by filling the pipe or container with inert gas, but this is a technique best left to an experienced welder.

SUMMARY

1. Silver soldering, despite its name, is a brazing process since it requires heat above 800°F.

2. Silver soldering is done exactly like brazing. The filler is melted at a temperature below that of the base metal and flows by capillary attraction into a gap (0.004 to 0.010).

3. The solder is melted by touching it to the hot metal. It is never melted by the flame.

4. Silver solder is an alloy of silver with metals such as cadmium, zinc, or copper. Gold is sometimes alloyed with the silver for use with precision instruments.

5. Fluxes used must be specially prepared for silver soldering.

6. Ordinary soldering is done at a flow temperature below 800° F. Solder is a mixture of lead and tin and sometimes traces of alloying metals.

7. Soldering is done exactly like brazing. The basic difference is the lower flow point of the binding material.

8. Solder is an excellent electrical conductor. This makes it ideal for jointing electrical connections where low heat and good conductivity are required.

9. Heat for soldering may be supplied by an oxy-acetylene torch, other gas torches, or by soldering irons.

10. Soldering fluxes are either acid type or rosin type. Rosin is used where a less corrosive flux is desired.

11. Sweating is a special soldering technique. The overlap area of two pieces of metal are tinned with solder. The tinned areas are placed face-to-face and heat applied on the outside to melt the solder between the lapped metal.

12. Soldered copper wire joints are common. One pipe fits inside the other with a 0.004 to 0.010 gap between them. Solder is flowed into this gap.

13. Stainless steel can be soldered, but it is difficult to handle. It requires a highly corrosive hydrochloric acid cleaner followed by the use of a zinc chloride flux.

14. Aluminum can be soldered but requires special solders. It is difficult and not recommended for beginners.

NEW WORDS AND TERMS

Rosin Flux A mild flux made of pine tar, used when a more corrosive flux is unsuited to a job.

Silver Soldering A brazing process that uses silver as the major constituent of the filler metal.

Soldering A process similar in technique to brazing but which uses filler material that melts below 800°F.

Sweating A soldering process. Two edges of metal are first coated with solder (tinning). The overlapped metal is then melted by applying heat through the lapped metal.

OXYACETYLENE FLAME CUTTING

Anyone who has labored with a hacksaw to cut steel will be amazed at how easily the job can be done with a flame cutting torch.

Flame cutting can be used on very thin sheet metal or very thick metal plate. There are even techniques that employ the oxygen lance to cut through metals which are several feet thick.

FLAME CUTTING TORCH

The flame cutting torch is basically the same as the welding torch, but with two exceptions. A center hole in the heating tip is connected to the oxygen supply. It is surrounded by several preheat holes (orifices). The operator can control the extra oxygen supply by means of a lever, which permits him to direct a blast of oxygen into the heated metal plate.

In ordinary fusion welding, the flame only melts the metal. In flame cutting the flame melts the metal in the same manner, but the blast of extra oxygen causes the metal to "burn" away. This extra oxygen blast is controlled by your hand and enables you to cut through metal just like a hand saw cuts through wood.

The actual cutting takes place through oxidation. Oxidation occurs naturally when metals are exposed to the oxygen in air over a period of time. In the case of iron and steel this oxidation—or combining of the metal with oxygen—is called rust, something we are all familiar with. You know that rust, if left to continue over a long period of time, will eventually eat away and destroy the metal. By directing a blast of oxygen onto hot metal we can accelerate the rusting process into an instantaneous destruction of ferrous metals.

The oxidized steel is blown from the cut in the form of a shower of hot sparks. The metal that is not consumed by the fire drops to the floor as molten slag. Slag is the oxidized form of metal, oxygen, and metal impurities.

Restrictions on Use of the Cutting Torch

Steel is easily cut with the cutting torch. Other metals, such as aluminum, stainless steel, and copper, to mention a few, are not as easy to cut as regular steel.

These other metals react unfavorably to the excess oxygen in the oxygen stream fed from the cutting torch.

Figure 13-1. The oxidized steel is blown from the cut in the form of a shower of hot sparks.

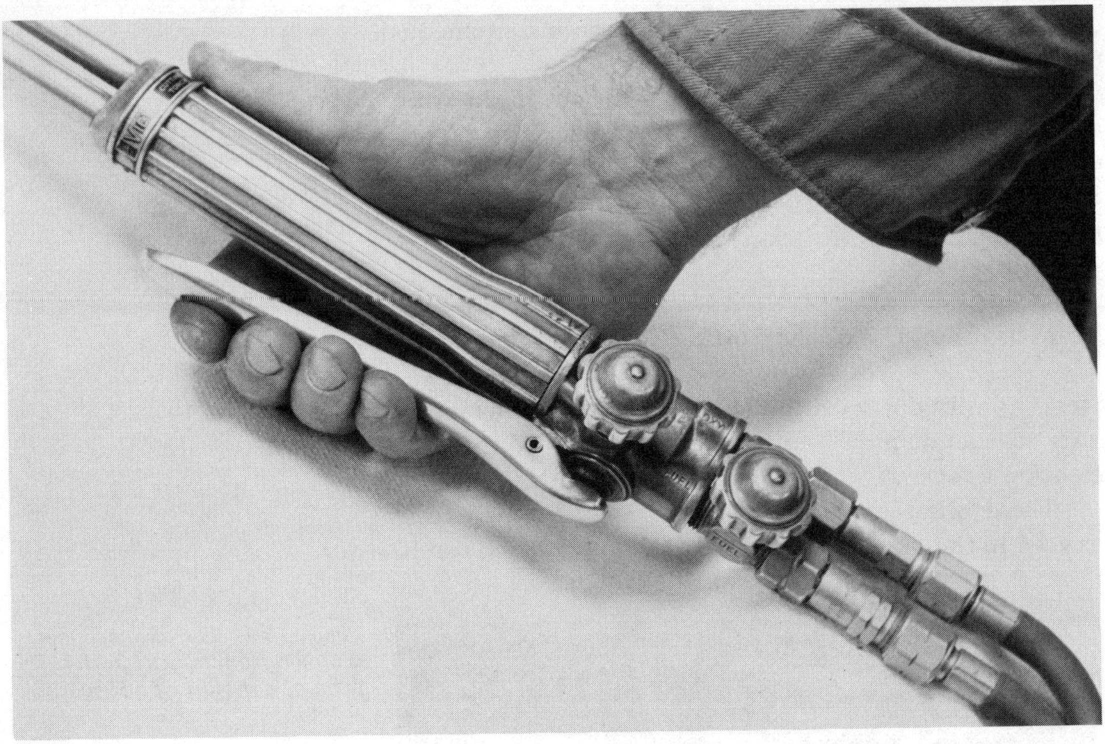

Figure 13-2. The oxygen blast for the cutting torch is controlled by a lever. In this particular model it is controlled by the fingers.

Therefore, they cannot be cut with the oxyacetylene flame cutting process. These metals can be cut with the powder cutting method, which will be explained later in this chapter.

Regulator

The cutting torch requires different settings on the oxygen and acetylene tank regulators than those we used for the welding torch. The settings are determined by the thickness of the metal to be cut. The following chart shows the regular settings for both oxygen and acetylene to use in cutting metals up to 2 in. thick.

Metal thickness, in.	*Oxygen pressure, lb*	*Acetylene pressure, lb*	*Hand cutting speed, in./min.*
1/8	7–20	3	20–30
1/4	25	5	17–26
3/8	30	5	15–24
1/2	40	5	14–22
3/4	40	5	13–20
1	50	5	11–18
2	50	5	7–13

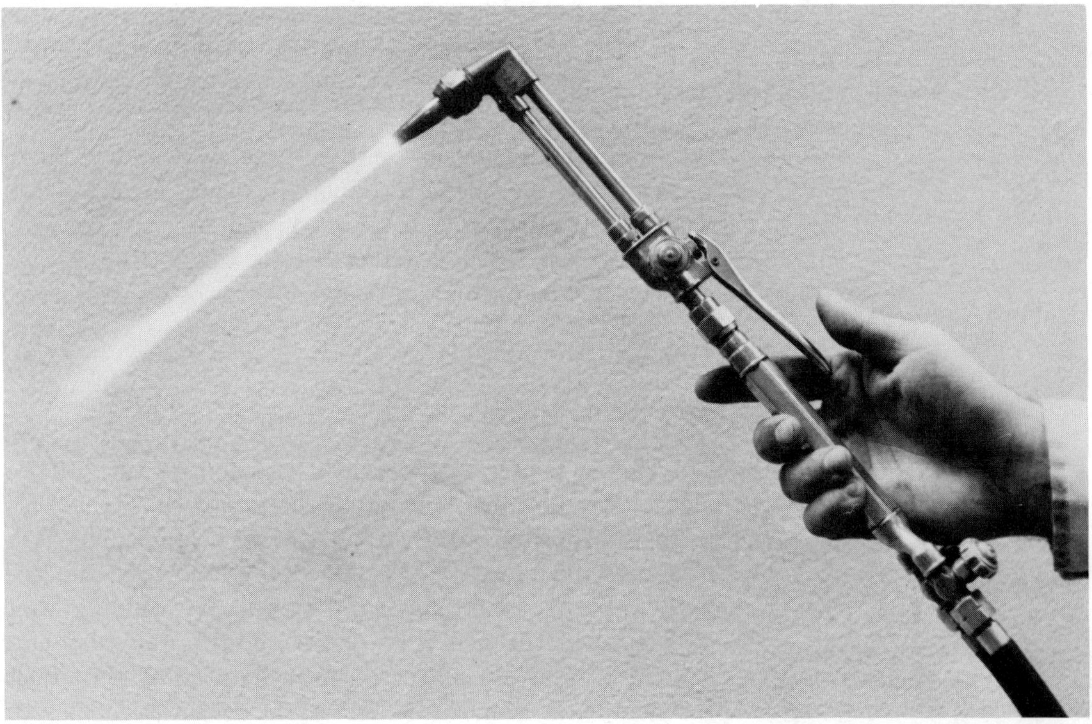

Figure 13-3. The cutting-torch attachment has a thumb-operated oxygen cutting stream lever. When the lever is depressed, a stream of oxygen blasts through the heated metal.

Oxyacetylene Cutting Torch

There are two types of oxyacetylene cutting torches: hand and automatic. The hand-held cutting torch is available as an individual item or as an attachment for the regular torch mixing chamber. The separate unit can be purchased with the oxygen control lever located on top for control by the thumb or on the bottom for control by the fingers. The finger control is shown in Figure 13-2 and the thumb control is shown in Figure 13-3. There is no difference in the use of the two. It is a matter of personal preference which one you select.

CUTTING TORCH VALVES

In the following discussion "cutting torch" will refer to the independent unit designed specifically for cutting. "Cutting attachment" will refer to the add-on unit that is used with the mixing chamber of a regular welding torch. It is screwed into place as an extension of the mixing chamber.

The flame cutting torch may have either three

needle valves to control the oxygen and acetylene flow or the regular two you are familiar with on the welding torch. The number of valves depends upon the manufacturer.

Two of the valves on the three-valve unit control oxygen and acetylene flow in the usual arrangement. The third valve adjusts the oxygen for the oxygen cutting stream. In the two-valve unit, the cutting oxygen is not adjusted or controlled except by the control lever.

Setting the Valves

The three-valve cutting torch is adjusted by turning the oxygen needle valve, located at the point where the hose joins the mixing chamber, all the way open. The two remaining valves—one oxygen and one acetylene—are used to control the flame type as in fusion welding; that is, first turn the acetylene valve about a half turn and use the striker to ignite the acetylene gas, then adjust the acetylene flame until the dark, sooty gas begins to

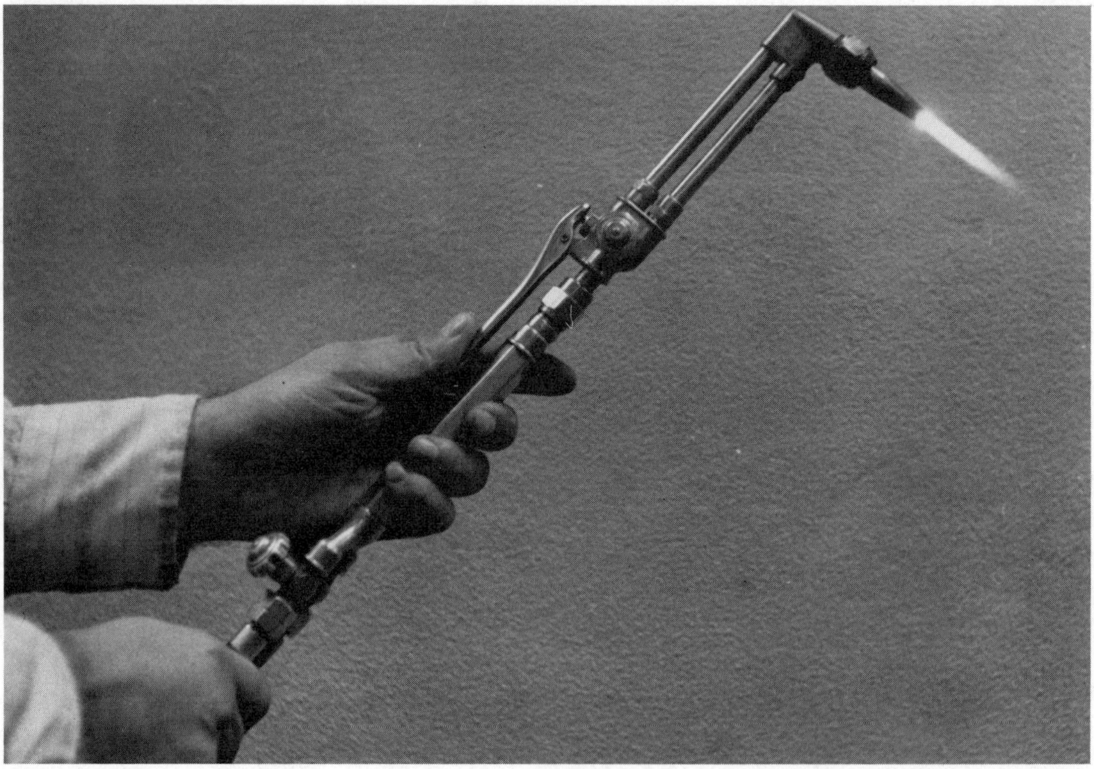

Figure 13-4. As the thumb depresses the oxygen lever, the torch flame is shortened. You will hear the noise of the oxygen blast when this is done.

burn clean. This will be the correct acetylene gas flow for the tip you are using.

Next, turn the oxygen valve until the flame turns to a neutral flame. Test the oxygen level to see if the oxygen circuit is working. When you depress the oxygen lever, there should be a blast of oxygen rushing through the center hole in the cutting tip. You can hear the noise and see a small clear area in the center of the flame as you check this circuit. As you hold down the oxygen stream lever, you should also check to make sure the inner heating flames remain in a neutral flame condition. Make any corrections with the mixture if this is not correct (Figure 13-4).

Changing Tips

The valve adjustments regulate the type of flame. In this case it is a neutral flame. But the temperature of the flame is controlled by the size of the tip opening just as it is in fusion welding. The tips are changed to increase or decrease the amount of heat to be obtained with a specific torch (Figure 13-5).

Figure 13-5. Tips for the cutting torch can be changed to obtain a hotter or cooler flame. Note the six preheat holes that circle the center hole. The center hole is for the oxygen cutting stream.

Several tip sizes are available for cutting torches and attachments. You determine the size tip according to the job you will be doing. Basically you must remember that the larger tip size will produce more heat and therefore will cut thicker metal. You should experiment with the torch on different thicknesses of metal to determine through experience the thicknesses that can be accurately and efficiently cut with different size tips.

Typical cutting tip sizes range from 00—the smallest—to 4. There are others, but this range is sufficient to cut metals from sheet metal thickness to 3-in. thick plate. This is all the range a beginner should attempt anyway.

Caring for Your Tips

Your cutting tips must be kept serviced and clean. Use a tip cleaner, as explained under welding torch care, to square the surface and clean the preheat holes and the oxygen hole.

Although the welding torch tips are only tightened hand tight, we use a wrench to tighten the cutting torch tips (Figure 13-6). Care should be taken that the screw end does not have scratches or dents on the mating

Figure 13-6. The cutting tip is being changed on this torch head. Different size tips allow you to control the flame temperature for cutting different thicknesses of metal.

surface where it joins the torch. Damage will prevent the cutting torch from sealing completely. In such cases a small flame may appear around the fitting. Such poor seating is usually caused by allowing the tips to roll around or drop on the floor. It can also be caused by not tightening the tip sufficiently into the mixing chamber.

PREPARING TO CUT

Remember that you are working with a tool that sends a shower of sparks for several feet in several directions. Make sure that the area is free of flammable materials that might ignite from the sparks. Also, keep yourself protected with gauntlet gloves, leather for your body, and a cutting helmet for full face protection.

It is a good idea to draw a line on the metal to guide you in cutting straight. This can be done with a piece of soapstone. It leaves a line that will remain visible in the intense heat given off during the cutting process. A piece of metal clamped in position along the projected cut can also be used to brace the cutting torch and keep you going in a straight line.

Start practice with metal plate between 1/8 and 1/2 in. thick. Set the plate so that about 3 or 4 in. hangs over the edge of the cutting table. Practice making the cutting stroke with an unlighted torch to make sure

Figure 13-7. Soapstone is used to leave a mark that remains visible during welding or cutting.

Figure 13-8. The torch must be braced to prevent uneven cuts. Here the gas pipes are braced by sliding the torch along the edge of the welder's arm. The small flame without a shower of sparks indicates that this is a preheat flame.

there are no restrictions in your way. You can ruin a cut by starting and hitting the clamp holding the metal when you are halfway through.

Steadying the Torch

You can rest your hand on the length of pipe used to mix the gases (the part of the torch between the tip and the mixing chamber) and push or pull to make the cut. The pipe passageways get warm but not hot and make an excellent brace to steady and guide your work (Figure 13-8).

For shorter cuts up to 6 in., another method of steadying your torch can be used. Here you brace the torch pipe against your fist to roll the torch from side to side. A slight adjustment for torch tip height must be made as this is done.

CUTTING METAL PLATE

The following procedures are used for cutting heavy metal:

1. Hold the torch tip at a right angle to the metal plate surface. This is for metal plate about 1/2 in.

or less. Thicker plate will have to be started with the torch held at a steeper angle on the edge and then, as the cutting stream has been established, angle the torch head into the 90° angle.

2. The torch is held about 1/16 in. above the plate.
3. Make sure your torch tip is lined up with the soapstone line and that your clearance is good for the length of the cut.
4. Use your preheat flame to heat the metal. You will notice small sparks jumping from the surface of the metal as you approach the melting point.
5. When the metal is bright red, slowly depress the oxygen lever to start the cutting oxygen stream through the center hole of the tip.
6. The oxygen blast into the bright hot metal will instantly burn up the metal, leaving a pathway or cut through the metal where the oxygen stream was directed.
7. This pathway or cut is called a *kerf*.
8. When you see a shower of sparks shooting down from the *underneath* side of the plate, it is a sign that the kerf has been established completely through the plate.
9. As you continue along the soapstone line, maintain the 1/16 in. height of the preheat flame's inner cone above the metal surface, and control your rate of forward movement.
10. The rate of forward movement is given in the table on page 196. For the 1/2-in. metal we are practicing on, the forward speed will be between 14 and 22 in./min. You can estimate the proper speed by watching the kerf zone to see if it is cutting all the way through the metal.
11. The cut should not be too slow. It must be fast enough to keep the sides from melting. If the speed is too slow, the sides of the kerf will melt and fill in behind the cut.

There will always be a small amount of back filling, but this is very oxidized metal that is so brittle it can be broken free with a sharp tap of a hammer. Caution: never use the torch tip to break this piece loose. You will ruin a good tip.

Effect of Too-Fast a Speed

If your rate of torch travel is too fast, the oxygen stream will not be able to penetrate through the metal. You will lose the kerf started by the oxygen stream and the cut will not penetrate completely through the metal plate. When this happens, you must go back to where the full cut stopped and restart the cutting. Use the same procedure you did to begin the cut at the metal's edge.

When starting or restarting a cut it is important to depress the oxygen cutting stream lever *slowly*. This is to prevent a sudden blast of cold oxygen from suddenly cooling the red-hot metal that has been brought to cutting temperature by the preheat flame.

If you do cool the metal, you will have what welders call a "cold start." Your kerf will be wider and more jagged than it should be.

Sign of a Good Cut

The overall objective of a good welder should be to make the cut as smooth as possible with the smallest kerf. He should make the cut in one continuous operation if he can. Otherwise, if he starts and stops several times, the kerf will be wider than necessary.

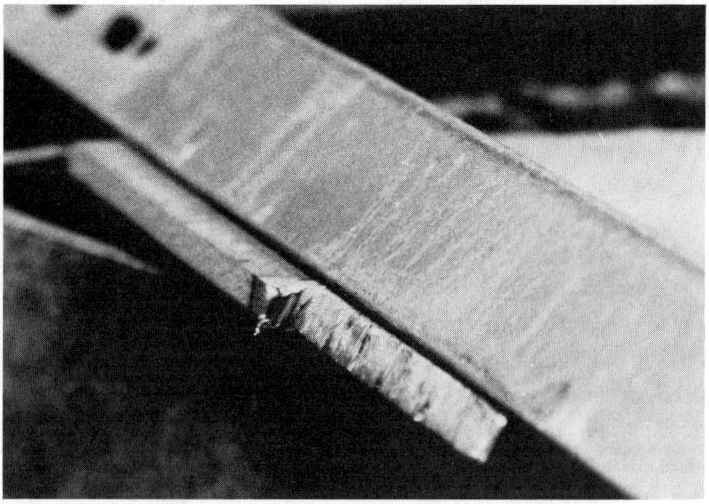

Figure 13-9. A smooth kerf is a sign of a correct cutting technique.

A wide kerf is not necessarily the sign of a poor cut, of course. Thick metals will normally have a wider kerf than thinner metals. This is because we must use a larger tip to produce sufficient temperature to cut the thicker metal. The bigger tip produces a wider flame.

Checking Your Welding Controls

Control, as used here, does not mean adjusting regulators and valves—it is *control of your operations* and involves the following objectives:

1. Selecting the correct tip size
2. Achieving the right preheat and starting procedures
3. Holding the torch at the correct angle to the metal surface
4. Having the preheat flame at the right height above the metal
5. Maintaining the correct rate of forward speed

Ending the Cut

Continue the cut to the end of the soapstone line. When the cut is completed, release the oxygen lever. Shut down the torch by closing the acetylene valve *first* as explained in the chapter on the acetylene torch, then turn off the oxygen valve or valves. If you have finished for the work period, release pressure in the lines in accordance with your standard safety procedures.

Breaking the Slag

When the cut is completed, you may be surprised to find that the cut portion is still stuck to the main body of the plate. Try tapping the cut piece lightly with a hammer. If it is held by slag, the brittle oxidized metal will break away cleanly. If it does not break off readily, then your penetration was insufficient and you have solid metal still holding the two pieces together (Figure 13-10).

Continue your practice until your cuts separate cleanly and the kerf zone shows a good appearance.

Figure 13-10. The slag formed at the root of the kerf area will hold together the two pieces. A slight rap with a hammer will break apart the brittle slag.

BEVELING

Beveling, so necessary to making fusion welds on thick metals, can be easily accomplished with the cutting torch. It is not only much faster than using a grinder, but it can also be done in places where you cannot get a grinder.

The bevel cut is done in the same manner as the straight cut. The torch is held at an angle to the base metal. The angle depends upon the requirement for the job.

It is important to remember that the area to cut will be longer since we are now cutting at an angle instead of straight through the metal. This may require using a higher temperature (larger) torch tip. Cutting speed may also have to be adjusted for the thicker metal.

Procedure for Beveling

1. Set up the metal plate with 3 or 4 in. hanging over the edge of the metal cutting table.
2. A piece of angle iron turned upside down will form a brace to help you maintain a 45° angle of the torch. This iron not only serves as a guide but helps you steady the torch (Figure 13-11).

Figure 13-11. A piece of upside down angle iron is used to guide the torch to cut a bevel on this piece of metal. The torch is preheating the edge of the plate.

3. The three important points in beveling are to keep the torch steady, to control the forward speed, and to keep the angle constant.
4. Start the cut by heating the edge of the metal plate until it is a bright red. Then, as you did in the straight cut, *slowly* depress the oxygen cutting lever to blast away the oxidized metal.
5. The oxygen blast will cause the metal to burn up in a shower of sparks.
6. Carefully control the rate of torch speed, keep the torch steady, and hold the torch angle and height correctly.

Figure 13-12. The oxygen blast causes the metal to burn up in a shower of sparks. Note how the angle iron is used to guide the angle of this bevel cut.

7. Watch in the kerf to determine if your forward speed is too fast or too slow.

Drag

Figure 13-13 shows the effect of improper forward speed where the torch was moved too fast. The speed was not too fast to prevent a complete cut, but the too-rapid movement of the oxygen stream caused the bottom of the cut to lag behind the top of the cut. The oxygen stream formed swirl or drag lines. This *drag*, or *lag*, as it is also called, results in curved lines along the sides of the kerf area.

SUMMARY

1. The capability of a cutting torch to cut steel is due to the property metal has to burn intensely in the presence of oxygen.

2. When steel is brought to a bright red heat and a blast of pure oxygen is blown upon it, the steel will oxidize rapidly, that is, burn.

3. The cutting torch is specially built to provide a means of directing a pure blast of oxygen into the heated metal. An extra pipe connects to the mixing chamber and is controlled by a lever operated by the welder's fingers or thumb.

Figure 13-13. An inspection of this torch-cut kerf zone shows that the cutting rate was too fast. Penetration was complete, but the drag lines on the inside wall show a curve to their form.

4. Pure oxygen is directed to the hot metal through a hole in the center of the cutting tip. Six (or four) smaller holes around the center hole provide the regular oxy-acetylene preheat flame to heat the metal to its cutting temperature (Figure 13-14).

5. A neutral flame is used in cutting and the regulators are adjusted to give proper pressure. Regulation is made according to the thickness of the metal.

6. Metal thickness also determines the size of the cutting tip. The larger tips produce more heat, which is necessary to cut thicker plates.

7. The metal to be cut is first preheated by the torch, then the actual cut is made by depressing the hand control to release a blast of oxygen into the hot metal.

8. The important technical points in making a good cut are to select the right tip, observe the correct preheat and starting procedure, hold the torch at the correct angle and height to the base metal, and maintain the correct forward speed.

9. Slag may form. Slag is oxidized metal. It is brittle and can be easily broken off.

10. The cutting torch is a quick way to bevel metal for fusion welding. Beveling is done in the same manner as straight cutting, except that the torch is held at an angle to the base metal. A piece of angle iron turned upside down provides a good angle reference and guide for making a bevel cut.

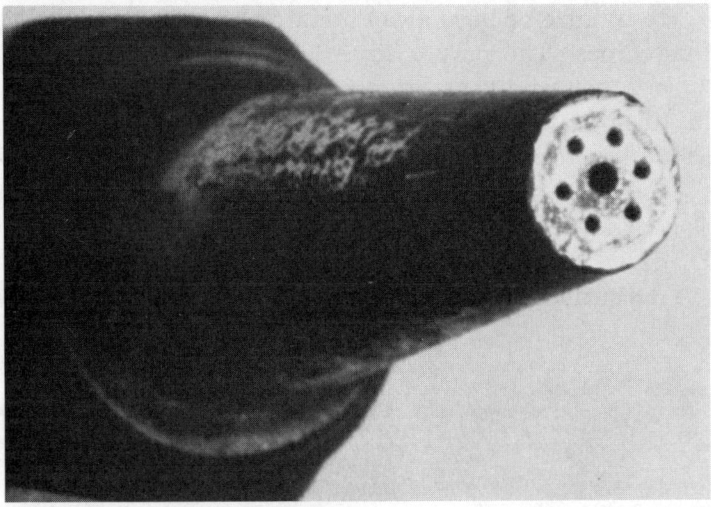

Figure 13-14. The center hole is for the oxygen blast controlled by the lever on the cutting torch. The six small holes provide the normal oxyacetylene flame that preheats the metal to the cutting temperature.

Figure 13-15. In beveling, the torch is held at an angle. An angle iron turned upside down is an excellent guide.

NEW WORDS AND TERMS

Cutting Attachment A special attachment added to the end of the mixing chamber of an acetylene fusion welding torch to permit flame cutting.

Cutting Tip Special tips used on the cutting torch. Whereas fusion welding tips have a single hole or orifice, the cutting torch has a single large hole in the center for the oxygen blast and six (or four) smaller holes surrounding the center orifice for the regular oxyacetylene preheat flame.

Cutting Torch A special oxyacetylene torch used for rapid cutting of metals. A hand-operated lever directs a blast of pure oxygen into preheated metal to oxidize or burn away metal in the cut.

Drag Lines The curved lines left on the face of torch cut metals by a too-rapid movement of the torch during cutting.

Kerf The term used to describe the path left by cutting action of the torch.

Slag The weak, brittle combination of metal and oxygen left after cutting or welding. It may form on the surface or the edges.

SPECIAL CUTTING TECHNIQUES

In Chapter 13 our initial introduction to flame cutting was with 1/2-in. plate. Metal thicker than 1/2-in. is cut in the same manner, but we start differently and the torch movement is slower.

CUTTING HEAVY PLATE

It is necessary to heat the edge of heavy metal at an angle to start the cut. After the edge is heated, the oxygen stream is added to begin the cutting. Begin by holding the torch at an angle along the edge. This permits heat penetration to spread more easily. Then slowly angle the torch tip up until it is in the right angle (90°) position we used in making a straight cut (Figure 14-1).

When metal thickness exceeds 1 in., it helps to preheat the edge before the cut is started.

Your cutting speed is very important. If you attempt to go too fast, you will lose your cut. A shower of sparks (the oxidized metal) will shoot up toward you instead of penetrating through the kerf and going downward. If this happens, restart the cut in the usual manner.

CUTTING SHEET METAL

Automotive repair requires a great deal of sheet metal cutting. The cutting torch makes short work of removing mangled hunks of wrecked car bodies.

Figure 14-1. The torch tip is angled up to 90° for making a straight cut on heavy steel plate.

Cutting sheet metal—metal less than 1/8 in. thick—is faster than cutting metal plate. You start the sheet metal cut with a torch tip set at a 90° right angle to the metal surface. Preheat the edge of the cut and then slowly depress the oxygen stream control lever. As soon as the metal is blown from the initial cut area, slowly tilt the torch angle to point in the direction of the cut. The force of the oxygen stream will actually blow the cut ahead of the preheat flame. You must practice this type of cut until you are able to keep up with the speed of the kerf. The angle of the torch tip to the metal surface is approximately 10° to 20° (Figure 14-2).

CUTTING HOLES

Your cutting torch can also be used to cut holes and circles. Cutting small holes is a handy way to make bolt holes in metal. The small holes can be made very accurately. Rest the torch mixing tube on your fist to hold it in position and slowly preheat the metal until it is bright red. Then slowly add oxygen as you raise the torch at the same time. Raising the torch slightly will prevent molten metal from clogging the torch tip as it bubbles to the surface.

The molten metal will bubble in small amounts until

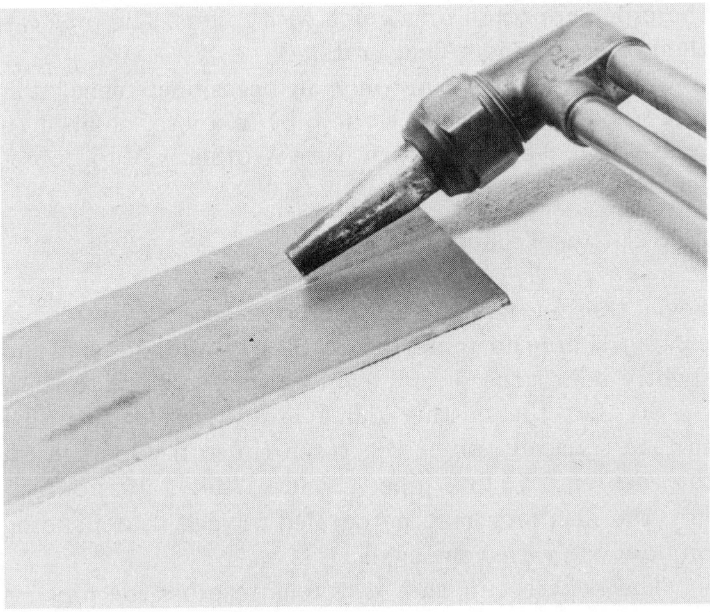

Figure 14-2. The cutting torch is shown here in the position used to cut thin sheet metal. Notice the angle of the torch head to the metal surface: about 10 to 20°.

the oxygen stream pierces the metal plate. When the metal is pierced, slowly lower the torch tip back to the regular height of 1/16 in. above the metal surface.

After the pinpoint hole has been made, carefully move the torch tip around the soapstone line you previously marked to map the dimensions of the desired circle or hole. The accuracy of the cut will depend on the steadiness of your control in following the line.

This procedure can be used for making pinpoint holes or larger circles. It can also be used for making square or rectangular holes. The starting procedure will always be to preheat a pinpoint area and then proceed with the rest of the cut.

PIPE CUTTING

Fabricating pipe systems often requires many joints and angles, which can be very conveniently cut with a torch. Likewise, if a pipe is broken, it can often be cut out with a torch and a new section welded in easier than the system can be dismantled and new pipe threaded in. The cutting torch can also be used for beveling pipe to make fusion welds.

Figure 14-3. Cutting bolt holes to exact diameter can be made by spiraling out from the center. This gives several revolutions to "true" up the hole.

There are two ways to set up pipe for cutting. If you will be using the torch on a lot of pipe, it is best to have the pipe supported on a guide with rollers. The pipe can then be easily and quickly rotated.

If you are handling only an occasional piece, it is more economical to lock the pipe in a vise, chain it to braces, put it in clamps, or use any other way to hold it securely.

Pipe Cutting Technique

Begin the cut with a small pinpoint area as you did in cutting a hole in plate metal. Slowly add oxygen and slightly adjust the torch height upward to begin the cut.

So that the cutting flame always meets the pipe surface squarely, point the torch tip so it aims toward the center of the pipe (Figure 14-4). If you do not, the kerf area may be beveled or jagged, depending on how you varied the angle.

The square cut—that is, straight across the pipe—is started at the top and brought down to the bottom of the pipe. Rotate the pipe, begin where the previous cut ended, and work down again.

BEVELING PIPE

Beveling pipe is one of the trickier cutting problems. It is done in the same manner as beveling flat stock, but

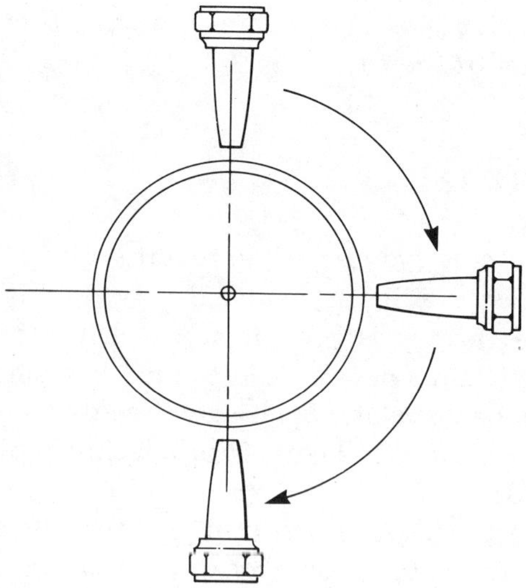

Figure 14-4. Position of torch tip during pipe cutting.

the problem is complicated by the curvature of the pipe. You must maintain the same angle of cut as you move around the pipe. It is an excellent exercise in freehand control. It requires great steadiness and constant speed of the cutting torch.

CUTTING RIVETS AND BOLT HEADS

It is often necessary in dismantling or repairing heavy machinery to remove rivets and bolts. This is no problem with the cutting torch. Most rivets and bolt heads extend above the surface of the metal they join. A special curved torch tip—called the scarfing tip—directs the cutting flame parallel to the base metal. This permits you to burn away the bolt head or rivet without digging into the metal (Figure 14-5).

To cut a rivet, preheat the rivet head until it is a bright red. Direct the oxygen stream into the rivet head and slowly cut away half of the head. Move the flame back to reheat the second half of the head. Again slowly apply the oxygen and cut away the remaining section of the rivet. A bolt head is handled in the same manner.

After the head is removed, a pin punch or similar object can be used to tap the shank of the bolt or rivet out of the hole.

Figure 14-5. The scarfing tip directs the flame parallel to the base metal, permitting the torch to cut away rivets and bolt heads without digging into the metal.

CUTTING COUNTERSUNK HEADS

Countersunk heads are those that are flush with the surface of the metal. They are more difficult to remove without enlarging the hole, which is undesirable, so we use a special low-velocity cutting tip to dig the edges away from the main rivet body. The low-velocity stream of oxygen keeps the tip from digging too deeply into the body of the metal. After the rivet head is removed, the bolt shank is punched out.

CUTTING LARGE CIRCLES

Cutting circles in metal requires a steady hand or a guide to control the accuracy of the circle cut. The actual cutting technique is no different from following a straight cut. Draw the required circle with soapstone and follow the line with a steady hand (Figure 14-6).

Special guides are available to control not only the radius of the circle but also the height of the torch tip. This leaves the operator with only the torch speed to control.

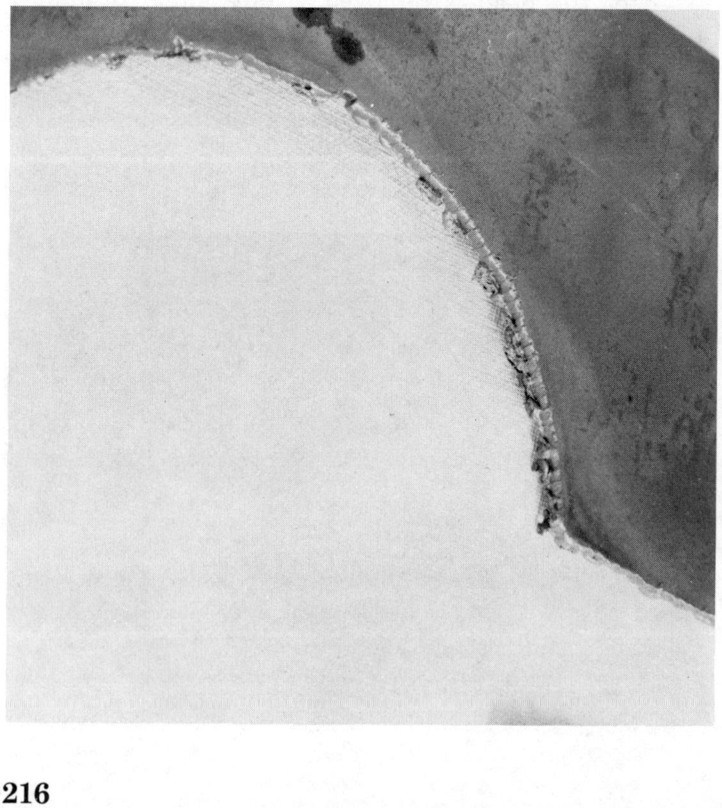

Figure 14-6. The cutting torch was used here to cut a circle into a piece of metal plate. Notice the slag deposit on the lower edge.

AUTOMATIC FLAME CUTTING MACHINES

Automatic flame cutting machines are available. They can cut steel with greater accuracy and workmanship than can an individual operator working manually. Some of these machines are portable and can be carried to material that cannot be conveniently brought into the welding shop.

Basically, the cutting torch operation remains the same, with the exception that an operator is not required after the cut has started. The manual portion of the work is then controlled by the machine.

Some machines are guided by a pattern that is traced by a stylus arm. The arm follows the contour of the pattern and then directs the cutting torch to make the proper cut. These machines can be designed to operate more than one cutting torch, which reduces the manufacturing cost by reducing production time.

Types of Flame Cutting Machines

The simplest form of automatic flame cutting machine is a motor-driven torch mounted to run on a track (Figure 14-7). The work material is located on a cutting

Figure 14-7. This automatic cutting torch is motor driven on a track. Notice the cutting torch assembly. It is fastened to the motor.

bed or table. This permits the cutting torch to travel across the material to make the cut along a designated line. You cannot make irregular cuts with this type of machine. It can make straight, square, rectangular, beveled, or circle cuts with a special adaptor.

The torch cutting speed is adjustable with a variable-speed motor. The manufacturer's manual gives recommendations for the proper cutting speed. These recommendations will have to be modified to compensate for different degrees of metal grades, tip wear, beveling, or machine motor differences. If these qualividations are observed, machines will give full penetration of the cut with a minimum of drag lines.

Operation and Setup of Machine Flame Cutters

The initial setup of a machine flame cutter is much like preparing to make a cut with a manual torch. You connect the necessary oxygen and acetylene hoses and set the regulators according to the recommended pressures for the metal thickness.

Next check the motor track to make sure there are no metal chips or slag splatter that might interfere with smooth motor travel. You must also check the full length of the cut area to make sure that the material is properly positioned in the cutting bed. It is necessary to also make sure that the length of the hoses feeding the motor are long enough. Otherwise the cutting will stop before you reach the end.

Check the manufacturer's recommendations for the cutting speed and set the speed regulator on the flame cutter controls. The usual controls for the automatic flame cutting machine are (1) a forward, neutral, and backward switch; (2) a speed range control; and (3) a speed range control within one of the ranges selected on the second switch.

Procedure for Operating the Machine

To start the cut follow these procedures:

1. With the motor switch turned *off*, manually position the torch tip near the edge where the cut will begin.

Figure 14-8. *This machine is used to control the automatic flame-cutting method. Notice the three controls: (1) forward, neutral, and reverse; (2) speed range; and (3) speed control within each range. These three controls allow the operator to select the proper speed of travel.*

2. Open the acetylene and oxygen needle valves and adjust the regulator pressure in the same manner as for manual welding.
3. Use a striker to light the flame; adjust the torch to give a neutral flame.
4. Manually move the torch tip next to the edge of the metal plate to be cut and allow the flame to preheat the edge.
5. The inner cone of the preheat flame should be about 1/16 in. above the surface of the metal.
6. When the metal turns a bright red, turn on the oxygen stream and the motor switch. Watch the kerf area and make any necessary adjustments to the torch cutting speed to control drag lines and any fusing of the cut.
7. Continue the preheat flame height as the torch travels along the track. Some warpage or buckling may cause the metal to move closer to the torch.

The height should be readjusted to the normal 1/16 in. This can be easily done while the torch is in motion.

Checking the Quality of the Cut

The automatic flame cutting machine should produce a much smoother and more accurate cut than manual cutting. You can compare the kerf area of a machine flame cut edge with a hand flame cut edge. The machine cut edge should be as smooth as if it had been milled, hacksawed, or machined. The hand cut edge will show the small variations due to unsteadiness and varying cutting speed, producing a less smooth edge.

POWDER CUTTING

Certain metals are not suitable for cutting with the oxyacetylene process. Stainless steels, cast iron, and aluminums are among these. They either build up what is known as refractory oxides that resist the oxygen process or the reaction of the oxygen to the metals tends to cool the cutting zone. This makes it difficult to start and maintain the cut.

The difficulty of cutting these and other high-alloyed metals led to the development of the iron powder cutting process. It operates as an addition to the oxyacetylene cutting torch. In the powder cutting process, a finely ground iron powder is added to the oxygen cutting stream. The powder becomes highly flammable as it is heated and blasted into the preheating flame. The effect of the iron powder is to maintain a higher degree of heat. The higher heat stabilizes the flame cutting and prevents the refractory oxides. This is because the iron powder balances out the excess oxygen that may form as these difficult-to-cut metals are heated to a cutting temperature.

How Iron Powder Is Used

The powder cutting torch requires a special means of adding the powder into the oxygen cutting stream. This is done either by an air pressurized system or a vibration dispenser.

The pressurized system uses several pounds of iron powder that is pressurized and blown into the cutting stream. This powder may be added either into the torch cutting head or through a tube connected to the outside of the torch cutting tip.

The vibrator dispenser allows the iron powder to go to the torch cutting head. Approximately 1/4 to 1/2 lb per min. of iron powder is blended with the oxygen cutting stream.

A powder cut may have a deposit laid against the metal. This is just an accumulation of powder. It is easily removed with a chipping hammer.

To begin an iron powder cut, depress the oxygen lever, which will deposit a small amount of iron powder ahead of the oxygen stream. The excess powder will raise the flame temperature quickly, making a preheat flame unnecessary. The cutting action can start immediately.

Raise the torch head about 1 to 1 1/2 in. above the surface of the metal. The cut then proceeds as normal. It may be necessary to use a tip size one size larger than normal on metal larger than 1 in. thick.

SUMMARY

1. Cutting plate over 1 in. thick requires preheating the edge before the cut begins.

2. Sheet metal—metal 1/8 in. thick or less—is cut with the torch tilted in the direction of the cut. The angle of the torch tip to the metal, after the initial preheating with the torch at a 90° angle, is approximately 10° to 20°.

3. Holes for bolts and rivets can be flame cut by first cutting a pinpoint hole and enlarging it to the desired size.

4. Pipe cutting is best done by rotating the pipe in a roller. The torch is held so that it always points in a 90° angle to the center of the pipe.

5. If no roller is available, lock the pipe in a vise or clamp. Cut from the top to the bottom on one side. Rotate the pipe and then cut the remaining side in the same manner.

6. Rivets and bolt heads can be removed by using the scarfing tip. This directs the torch flame parallel to the metal, minimizing dig in.

7. Circles are cut by first drawing a pattern with soapstone and then following the pattern line with a steady hand. The technique is like that of a straight cut.

8. Automatic flame cutting machines produce cuts with greater accuracy and workmanship.

9. Controls for automatic flame cutting machines include a switch for forward, neutral, and reverse; a speed control range; and a speed control within each of the ranges.

10. Speed is set according to the manufacturer's recommendations and then adjusted as visual inspection indicates.

11. Metals like stainless steels, cast iron, and aluminum are difficult to cut with an oxyacetylene torch. They require the iron powder process.

12. In the iron powder process, a finely ground iron powder is added to the oxygen stream. This prevents the refractory oxides that interfere with ordinary oxy-acetylene torch cutting of these high-alloyed metals.

NEW WORDS AND TERMS

Automatic Flame Cutting A machine used to flame cut metals with the oxyacetylene process.

Powder Cutting A method used to cut difficult metals such as stainless steel and aluminum alloys. Iron powder is introduced into the flame of an oxy-acetylene torch to control temperatures.

Refractory Oxides The build-up of oxides when heat is applied to stainless steel, the aluminums, and other metals. In some cases they act as a shield against the cutting stream. They are controlled by the iron powder cutting method.

ARC WELDING

PRINCIPLES OF ARC WELDING

Arc welding is older than oxyacetylene welding but did not come into general use as early. It works on the same principle as lightning bolts or the spark that jumps between the electrodes of an automobile's sparkplugs. This principle is that electrical current will generate large amounts of heat when it meets a heavy resistance. A resistance, in electrical terms, is something that is difficult for electricity to flow through.

ELECTRIC ARC

Electricity flows easily through some materials. Copper, gold, and water are three examples. These materials are said to be good conductors because they have a low resistance to electrical flow. Air, on the other hand, is a poor conductor. It offers great resistance to the flow of electricity. Therefore, we must put more voltage (pressure) behind the current to make it jump across an air gap between two electrodes (electrical connections). The resistance of the air to the current leaping the air gap creates great heat. This heat is expressed as a white-hot spark.

This is exactly what happens in an arc welding system. In its simplest form, we connect an electrical conductor to a suitable current. The end of this conductor is held close enough to a weld joint for the voltage to force the current to leap across the arc gap between the end of the conductor and the base metal to be welded. The resistance of the air between the conductor and the base metal creates heat ranging from 6000° to 10,000°F.

The amount of heat generated by this electrical arc is governed by the size of the conductor (the electrode) and the amperage (amount) of electricity available. By changing these two, we can change the amount of heat available to make our arc weld.

These principles were discovered in 1801, but no practical use for arc welding was found until 1887, when the electric arc was used to join lead plates in a lead-acid storage battery. The process was improved, but there was still a great deal of difficulty with the process. It was extremely difficult to control the arc stream when joining steel metals and there was no protection against the build-up of oxides in the weld puddle. This resulted in a weak and porous weld. In 1917 the stick electrode was developed and some fluxing was done by hand. Finally, in 1927 a coated or fluxed electrode was developed, leading to rapid advancement for arc welding because of its speed in joining thicker metals and the strength produced by arc welds.

Electrical Circuit

Electricity will not flow unless it has a complete circuit. It must return to the place it started. We provide this necessary circuit in arc welding in this manner:

1. The electricity is furnished by a welding machine. Welders call this an *arc welder.* The machine takes incoming electricity, usually 220 or 440 V, and changes it into the correct amperage (quantity) and voltage (pressure) needed for arc welding. The arc welder will step up (increase) the amperage and step down (reduce) the voltage.
2. This altered electricity (Figure 15-1) then flows from the machine (*A*), through the cable (*B*), to the electrode holder (*C*) gripped in the welder's hand.
3. The current then flows through the electrode gripped by the electrode holder. The electrode is a thin metal rod coated with special substances (Figure 15-2) that make it a good conductor. The electricity flows easily through it and leaps the air gap between the tip of the electrode and the base metal being welded. The distance across the gap is about equal to the diameter of the electrode.
4. The current then flows through the steel being welded and into the metal table that supports the

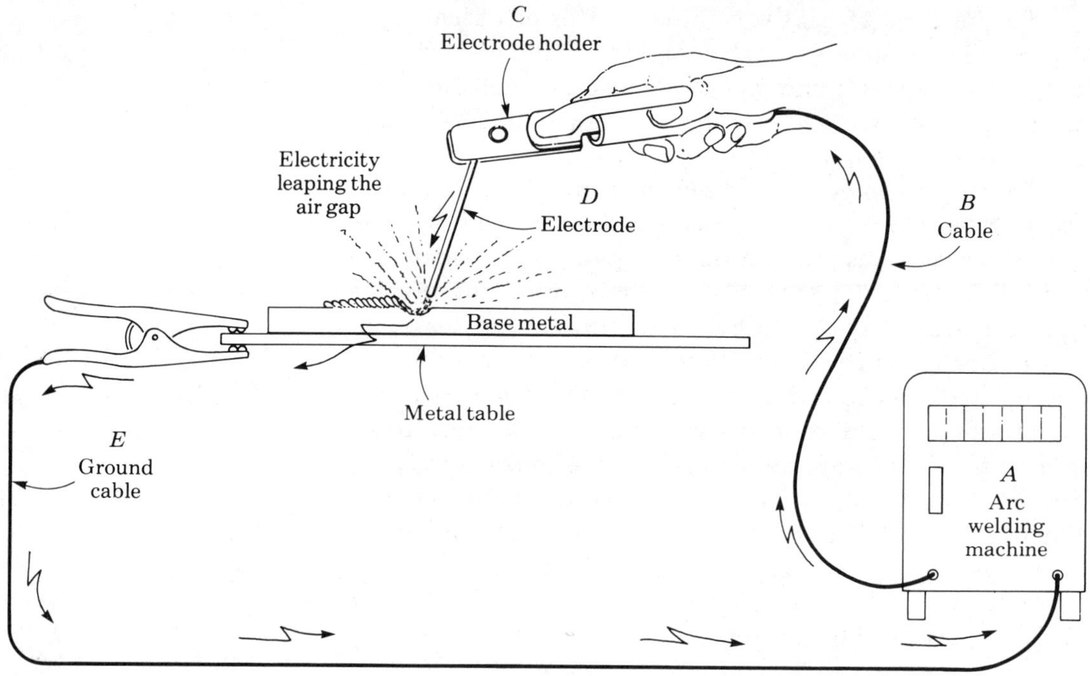

Figure 15-1. Electrical circuit in the arc welder.

Figure 15-2. Arc welding electrodes are coated with a chemical substance to aid the welding process.

work. A ground cable (*E* in Figure 15-2) clamped to the table (or to the work if no table is used) carries the current back to the arc welding machine to complete the required circuit.

5. Depending upon the polarity of the current, the current may also flow from the machine to the work and then up the electrode back to the arc welding machine. Polarity will be explained in a later section.

How the Arc Creates a Weld Bead

Figure 15-3 shows how the electric arc creates a weld bead. The electrode takes the place of both the oxyacetylene flame and the filler rod we used in oxyacetylene fusion welding. It is called a consumable electrode.

The arc produces sufficient heat to melt both the work metal and the tip of the electrode. The electrode and work metal melt together, and the flux on the rod creates a gas that shields the puddle from the air. Control of the arc puddle will be discussed in the next chapter.

The voltage produces pressure in the electric arc. The effect of this pressure on the molten base metal in the puddle is to "dig" out metal. This forms a hole or *crater* in the area melted by the arc. The pressurized "arc stream" pushes back the molten metal and melted electrode. As the weld puddle solidifies, this piled-up metal from the crater forms ripple ridges to create the

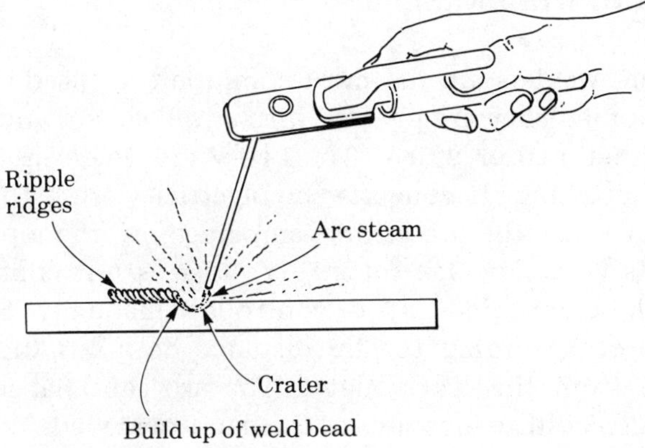

Figure 15-3.

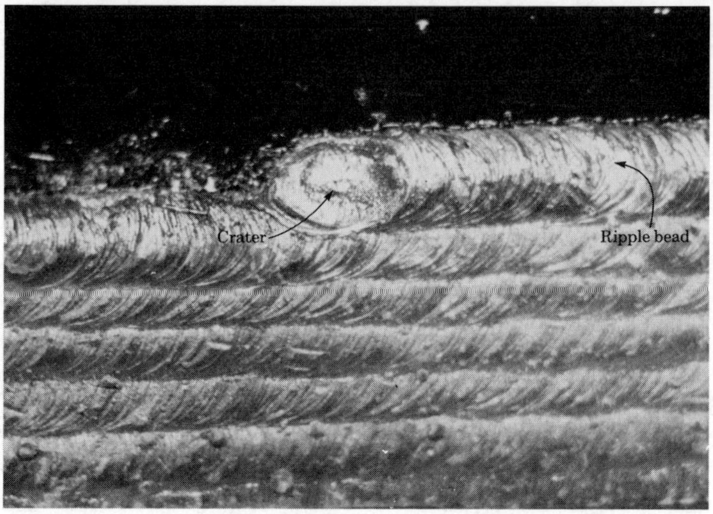

Figure 15-4. The force of the arc stream creates a "crater" in the steel. The molten metal and filler from the electrode is blown back to create the characteristic ripple effect of the weld bead.

weld bead. This bead looks a lot like the oxyacetylene weld bead, except that the ridges are heavier (Figure 15-4).

TYPES OF ELECTRICAL CURRENT

There are two types of electrical current: direct current (DC) and alternating current (AC). Both are used in welding. Direct current flows in one direction only. Alternating current changes direction according to its cycle. Most American current is 60 cycles per second. The current changes direction twice in each cycle or 120 times a second.

AC ARC WELDERS

AC arc welders are the most common type used around the home. They are plugged into a wall socket and work on either 110 or 220 V. The 110-V arc welder is a light-duty machine. It converts the electricity from the wall socket into the necessary amperage as the operator selects the amperage setting. Voltage is not controlled by these machines. It is controlled manually by the operator according to the distance he holds the electrode from the work metal. The arc gap changes the voltage. Voltage increases as the gap is widened.

Maximum voltage attainable is about 65 V. This occurs when the air gap is at its maximum. High voltage is not desirable for it digs too deeply into the metal.

DC WELDING MACHINE

A direct current arc welding machine may be either a motor-driven generator that produces DC current or a rectified AC arc welder. The rectified AC arc welder takes wall socket AC current and changes it into direct current. The DC arc welding machine may be driven by an internal combustion engine or an electric motor driven by AC current. Each type of machine has advantages and disadvantages. A major advantage of the gasoline-powered arc welder is that it is portable and can be used in areas where regular line current is not available.

POLARITY

Some machines are able to produce three separate electrical welding conditions: AC current, DC straight polarity, and DC reverse polarity.

Polarity of an arc welding machine may be compared to a person who is right or left handed. It refers to electrical poles. Direct current has a *positive* and a

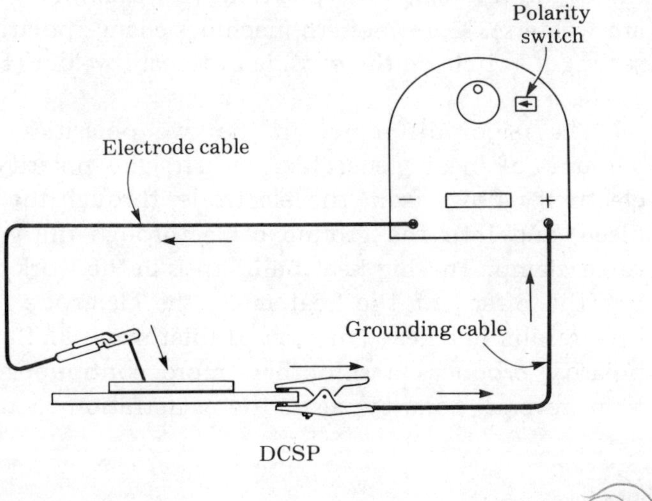

Figure 15-5. Direct current straight polarity (DCSP) in an arc welding machine.

Figure 15-6. Direct current reverse polarity (DCRP) in an arc welding machine.

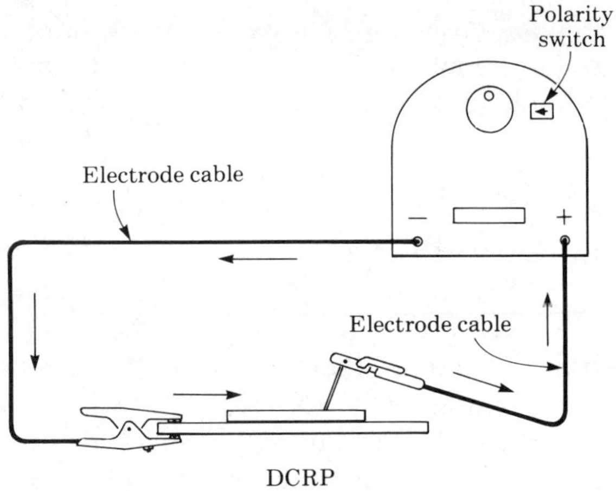

negative pole, just as the earth has a north and south pole. Normally, electricity flows from negative to positive. We call this *straight polarity*. It is abbreviated DCSP for DC current straight polarity. Straight polarity current flows from the electrode (negative) to the work piece, through the ground cable clamp (positive), and back to the arc welding machine.

In *reverse polarity* (abbreviated DCRP) the current flows through the ground cable clamp (negative) to the work piece, into the electrode (positive), and back to the arc welding machine.

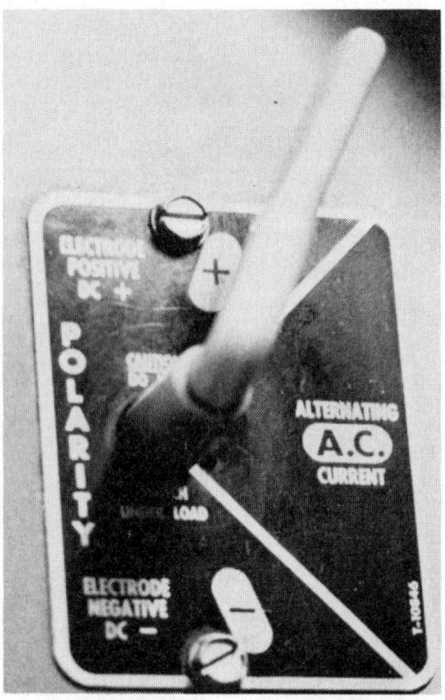

Figure 15-7. This is the polarity switch on one type of arc welder. The polarity can be changed from straight to reverse by moving the projecting handle. Or it can be used in the middle position as just an AC arc welder.

Changing Polarity

Polarity is changed on some machines by unplugging the cables and reversing their positions (for example, Forney arc welders). More modern machines change polarity by setting a switch on the outside of the arc welder (Figure 15-7).

The major difference in the two polarities is the amount of heat generated. In straight polarity the electricity flows from the electrode, through the work piece, and into the ground cable through the ground cable clamp. The big heat build-up is in the work piece, but the effect of the heat melts the electrode faster. This results in a heavy deposit of filler material. Straight polarity produces a smoother, more continuous bead than reverse polarity—also, the penetration is not as deep.

In reverse polarity the electricity flows in the opposite direction to straight polarity. This is desirable in making overhead and vertical welds for it enables the bead to freeze faster with less likelihood of sag or drip. It is also desirable in achieving deeper penetration through the base metal. Reverse polarity also helps to break down the oxides that form on nonferrous metals, such as copper and aluminum.

ARC LENGTH

The arc gap between the electrode and the base metal is called the *arc length*. The control of the arc gap is tricky for a beginner because the electrode grows increasingly shorter as it melts. This makes it necessary for the welder to "feed" the electrode into the weld puddle as it grows shorter, which means that the welder's hand must lower the electrode gradually as it is moved down the weld joint. This must be done at a rate that will keep the end of the electrode maintaining an even arc gap about the diameter of the electrode. This is difficult at first, but it becomes easier with practice.

Effects of Improper Arc Length

If the electrode is held too closely to the base metal (that is, the arc gap is too small), it creates a cold weld and may cause the electrode to stick to the base metal.

Too large an air gap creates too hot an arc. This causes the weld puddle to splatter and become uncontrollable because the voltage buildup is greater with a larger arc length. The greater heat produces a larger weld puddle.

If the arc length is about the diameter of the electrode (the recommended distance), the arc will produce a steady "frying" noise like bacon on a hot griddle. A trained ear can detect a noticeable difference in this noise as the arc length changes.

ARC BLOW

A disadvantage of the DC welder is that you may sometimes get *arc blow*. This is when the arc changes position even though you have a steady grip on the electrode holder. This condition occurs most frequently in corners and deep grooves. It is caused by electromagnetism that the electricity sets up in the base metal. This magnetism pulls the arc out of shape. If the deflection of the arc is opposite to the line of electrode travel, the condition is called *back blow*. If the arc blow is in the line of weld travel, it is called *forward blow*.

Arc blow causes a break in the evenness of the bead, leading to a weak weld. It also causes excessive splatter.

Counteracting Arc Blow

There are several ways to counteract arc blow. They are:

1. Reduce current.
2. Change position of ground.
3. Use as short an arc as possible.

If these remedies do not help, then you may have to shift from DC to AC current, which does not produce arc blow. Magnetic lines of force—the cause of electromagnetism forming in the work metal—do form around alternating current wires. But since the AC current shifts constantly from negative to positive, the current goes through the zero phase so often that the strength of the magnetic field is greatly reduced.

SAFETY PRECAUTIONS

High-voltage electricity is dangerous, but only if it is mishandled. The voltage coming into your welding machine—except small models designed to work on house current—have an electrical input of either 220 or 440 V. This is strong enough to kill a person. Therefore, the welding machine must be well grounded and all insulation checked at regular intervals. Any frayed or broken insulation must be replaced without delay.

You should especially check the tightness and insulation of the electrode holder and the cables. The cables must be of sufficient size to carry the current. If the wires are too small, they may overheat. The ground cable must have a firm grip. Any looseness may cause overheating or arcing (Figure 15-9).

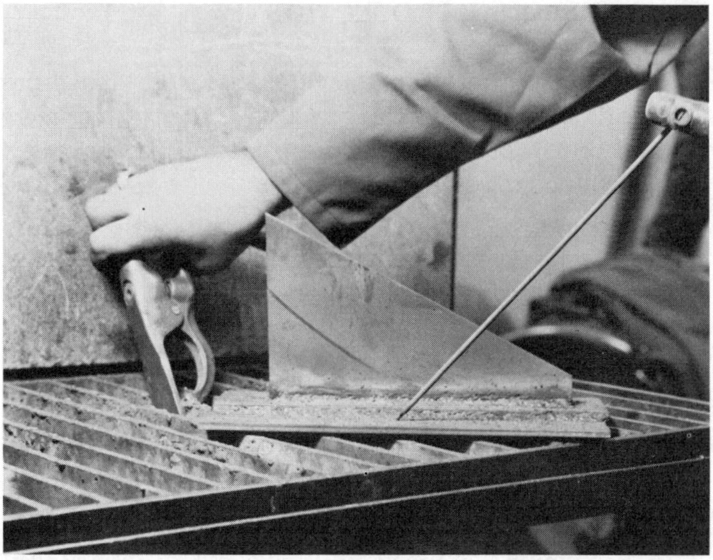

Figure 15-8. Sometimes moving the ground clamp farther from the weld joint will help reduce arc blow.

Figure 15-9. The ground cable must have a firm grip to complete the circuit and also to prevent arcing and overheating.

YOUR PERSONAL PROTECTION

Arc welding produces greater heat and more dangerous radiations than oxyacetylene welding. Safety rules were pointed out in an earlier chapter, but by way of review the following are essential:

1. Never look at an arc without eye protection. Intense ultraviolet and infrared rays can seriously damage your eyes. Always use a full shield.
2. The density of the filter in the head shield depends upon the amount of amperage used in the welding. Shade 14—the deepest—is used for heavy duty welding over 400 amps. Shade 12 is used for welding with an amperage of 200 to 400. Shade 10 is sufficient for up to 200 amps.
3. Arc welding produces more splatter. This requires clear glass lenses to protect the filter glass. The clear glass lens must be replaced when it becomes badly splattered.
4. Welding goggles are not sufficient for arc welding. They offer no protection for the face and neck. Use a helmet.
5. Leather gauntlet gloves are necessary to protect your hands from radiation and splatter.
6. Heavy leather clothing is required to protect your body from splatter and radiation that can burn through thin clothing.
7. Water is an excellent conductor of electricity. This makes it dangerous to arc weld in a damp or wet area. Keep the floor and work area dry.
8. Do not forget ventilation. More gases and fumes are created by the hotter arc. These must be removed. Adequate ventilation is a must. If working on locations where there are dead air areas, then you must set up portable fans (Figure 15-11).

SIMILARITIES TO OXYACETYLENE WELDING

Arc welding equipment, filler material, and application technique are different from oxyacetylene welding. However, many practices are identical, so if you have thoroughly familiarized yourself with the oxyacetylene process, you have learned three-fourths of what you need to know to be an arc welder.

Figure 15-10. Heavy protective clothing, preferably of leather, is required to protect the arc welder from excessive radiation and splatter.

Figure 15-11. These arc welding booths are properly vented at each work station by the pipe ducting supplied from a central air system.

The first similarity is preparation of metal. Clean the metal of rust and oxides. These, if left, will melt and fuse with the arc weld bead the same as they do with oxyacetylene, producing weak welds that may fail in service.

Bevel the plate before welding as in oxyacetylene welding. The arc has greater penetration than the oxyacetylene flame, but it still will not dig deep enough into the metal to give full penetration without beveling.

Use the same butt, T, lap, corner, and flange joints that you learned in the beginning. The basic welds are still the fillet and flat bead, and you will weld in the same flat, vertical, horizontal, and overhead positions. You will have to work a little harder to prevent sag and drip on the vertical and overhead welds because the arc is hotter than the oxyacetylene flame.

The actual technique of laying the weld bead is simpler because you do not have to dip and feed the filler rod into the molten crater. This is done for you by the melting electrode.

SUMMARY

Electricity creates great heat when it passes through a resistance. A resistance is anything difficult for the current to flow through.

1. When electrical current is forced by its voltage (pressure) to leap across an arc gap, heat is created by the dry air's resistance to the passage of the electricity.

2. The welding arc will range in temperature from 6000°F to 10,000°F, depending upon the size of the electrode and the amperage of the current.

3. The welding setup consists of an arc welding machine—called a welder—cables, an electrode holder, electrode, and ground cable.

4. Arc welding machines may be either AC, DC, or a combination of AC and DC.

5. The more versatile DC machine permits changing polarity.

6. In straight polarity the current flows from the negative terminal of the welder, through the work, and back to the positive terminal of the welder.

Figure 15-12. Proper clothing, like that hanging in this shop, and a clean place to work are essentials.

7. In reverse polarity the current flows from the welder through the work piece and back to the welder through the electrode, electrode holder, and cable.

8. Straight polarity deposits a heavier amount of filler, and reverse polarity produces deeper penetration.

9. High-voltage electricity is dangerous and must be treated with respect.

10. Proper clothing and a clean welding shop are essentials.

NEW WORDS AND TERMS

AC Electrical current in which the flow is first in one direction and then in the other.

Arc Electrical heat caused by resistance of electricity to air as the current leaps an air gap.

Arc Blow Deflection of an arc stream by magnetic forces.

Arc Length The distance between the base metal and the tip of the electrode.

Arc Stream The force created by the arc. It digs the crater and piles up molten metal to make the ripple ridges.

Amperage Measure of the amount of electricity in a current.

Conductor A substance that easily passes electrical current.

Crater A depression made in the molten puddle by the force of the arc stream.

DC Direct current that flows in one direction only.

Electrode Chemical-coated mild steel rod used in arc welding both as a conductor of current and a filler rod.

Electrode Holder Spring clamp attached to the welding cable. It holds the electrode.

Polarity The poles of an electrical current, consisting of a negative pole and a positive pole.

Reverse Polarity When the current flows through the work piece and back through the electrode holder to the welder.

Straight Polarity When the current flows from the welder through the electrode to the work piece and then through the ground to the welder.

Voltage Electrical pressure that moves electricity through a conductor.

ARC WELDING ELECTRODES

The ease and quality of your arc welding bead depends upon selecting the correct electrode. There are many to choose from, and they are divided into groups and codes.

ELECTRODE GROUPS

1. **Fast-freeze group (Codes E6011, E6010).** These are general-purpose electrodes designed to give deep penetration with deposits that harden quickly. The quick-freeze characteristic makes these electrodes good for overhead and vertical welds for they minimize sag and drip. They have little slag formation and produce a relatively flat bead. The two codes within this group (E6011 and E6010) perform similarly, but E6011 has the flux composition to enable it to be operated on AC current, and the E6010 does not.
2. **Fill-freeze group (Codes E6012, E6013, E7014).** This group produces a less penetrating arc than the previous group. They are commonly used in DC straight polarity, which provides a fast electrode melt and a higher deposit rate. They produce a slag covering. They can be used in all positions, but fast-freeze electrodes are preferred for overhead welding.
3. **Fast-fill group (E6027, E6024).** This group is recognized mainly by the heavy iron powder flux coating the electrodes. They produce smooth beads with heavy slag deposits that are self-cleaning. They are used in a flat down position only. These electrodes were developed in Germany and first

marketed in the United States in 1953. They have become very useful in production line work. These electrodes can be used in either AC or DC positions, and are applied with a drag technique. Weaving is not recommended.

4. **Low-hydrogen group (E7018, E7028).** This group is very important where special strength must be maintained in a weld. These electrodes are designed to have a minimum amount of moisture in the flux. This permits them to produce maximum strength in "problem steels." They have outstanding crack resistance and eliminate porosity in sulfur steels.

Figure 16-1. Electrodes are chosen by the information printed on the flux coating. The coding is used to represent the different classifications of electrode properties.

ELECTRODE CODING

The coding system used for electrodes is a standard jointly controlled by the American Welding Society (AWS) and the American Society for Testing Materials (ASTM). Individual companies may have their own codes for their products, but their shipping labels will make reference to the standard electrode identification code.

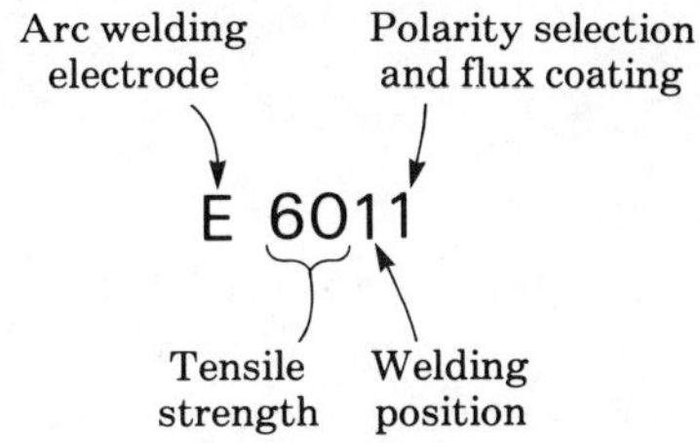

Figure 16-2. Manufacturers' electrode characteristics.

How the Coding System Works

The group classification gives the general characteristics of the electrode. The code gives specific information about it. It may consist of four or five numbers preceded by the letter E.

Suppose the code stamped on the electrode is E6011. The code then indicates the following:

E This letter identifies the electrode as being for arc welding use.

60 The first two numbers represents the tensile strength of the resulting weld bead in thousands of pounds per square inch. In this example 60 represents 60,000 lb of tensile strength per square inch of bead surface. *If there is a five-number series in the code, then the tensile strength number will have three figures instead of two.*

EXX*1*X In the four-digit code the third number represents the position the electrode can be used in for an adequate weld. Number 1 is for all positions, number 2 for either horizontal or flat positions, and number 3 is recommended only for flat position welding.

EXXX*1* This is the final digit in the code. It indicates the polarity and the flux coating.

Final Digit Double Meaning

Since the last number carries a double meaning, there has to be a further breakdown to distinguish which welding polarity and which flux may be used with that particular electrode. The breakdown is:

Code for Welding Current Polarity	Code for Flux Coating
0 DCRP when third digit is 1	0 Cellulose sodium
0 DCSP or AC when third digit is 2	
0 DC either polarity or AC for flat position.	
1 AC or DCRP	1 Cellulose potassium
2 DCSP or AC	2 Titania sodium
3 DCSP, DCRP or AC	3 Titania potassium
4 DC either polarity or AC	4 Iron powder titania
5 DCRP	5 Low hydrogen sodium
6 AC or DCRP (low-hydrogen potassium)	6 Low-hydrogen potassium
7 DCSP or AC for horizontal fillet or flat position welding.	7 Iron powder, iron oxide
8 DCSP or AC	8 Low hydrogen, iron powder

Company Codes

In addition to this numbering code, each company may market its own particular brand of electrodes under its own code, but the companies include a reference to the AWS classification.

Color Codes

Color coding is another popular method of electrode identification. This method has a colored dot on the "grip" end of the electrode. When the AWS code is printed on an electrode a color dot is not used. This is the case with E6010 electrodes.

Additional color coding is used to represent the end color and the flux color. A general-purpose electrode selection chart will give a great deal of information about the composition and use of a particular electrode. These charts will list for any given electrode:

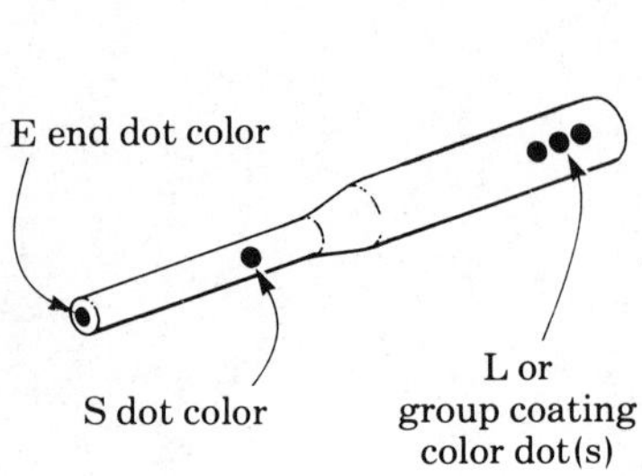

Figure 16-3. Color identification markings.

1. End color
2. Spot color (S dot)
3. Group color
4. Polarity selection
5. Amperage setting for various electrode diameters.

SELECTION CHARTS

Electrode selection charts are generally available from companies manufacturing electrodes. For example, Lincoln Electric's chart shows the color coating, including end mark, spot, comparative AWS code, company name, polarity, electrode sizes, and current ranges in amps. Of course, such data are general and welder experience must be used to modify it. Actual machine settings may have to be adjusted because of plate thickness, joint position, and operator preference.

ELECTRODE DIAMETER

Arc welding is generally done on metals ranging from 18 gauge sheet metal to much heavier plate metal several inches thick. You will most likely limit your beginning level arc welding to metals in the 1/4- to 1/2-in. thickness range.

Different thicknesses of metal require different heat inputs. We have noted how changing polarity changes heat input, but heat is also controlled by the diameter of the electrode. Electrode diameters vary from 1/16 in. to 1/4 in. This diameter is measured across the *bare* wire core of the electrode. Do not include the thickness of the flux coating in this measurement. With practice you will be able to easily detect a difference of 1/32 in. in the electrodes.

Changing electrode diameter regulates the amperage of the arc welding machine. This is similar in principle to changing tip sizes in oxyacetylene welding.

Electrode sizes increase in diameter by 1/32-in. graduations: 1/16, 3/32, 1/8, 5/32, 3/16, 7/32, and 1/4.

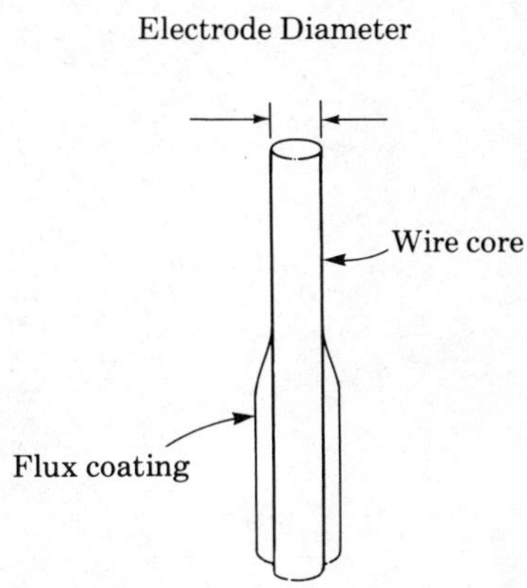

Figure 16-4.

AMPERAGE SETTINGS

There are recommended amperage range settings on the welding machine for each type of electrode. The electrode manufacturer or the welding machine manufacturer will list the suggested range of operation for a certain diameter of electrode. An E6011, 1/8-in. diameter electrode might be listed with an amperage range setting from 90 to 130 amps. You are given a range instead of a specific number because you will need a higher amperage setting for a 3/8-in. thick plate than for a 1/4-in. thick plate.

Of course, if you exceed the recommended range, the heat will be so great both electrode and body metal will melt excessively and the weld puddle will be uncontrollable. Excessive heat also produces excessive splatter. If your amperage setting is below the recommended range, the electrode will not start easily and it may stick frequently.

It is important to remember that amperage settings, current or polarity settings, type of electrode, position of electrode, type of metal being welded, and manipulation of the electrode all affect the ability of the welder to control the weld puddle and make a successful weld.

PENETRATION

The tensile strength of the electrode is given in the first two digits of a four-digit code and in the first three digits of a five-digit code. However, this tensile strength of the electrode is not all that determines the strength of the weld bead produced by this electrode. Penetration of the weld bead is a most important factor in securing a strong weld joint. Penetration depends upon several factors. These include amperage setting of the machine, joint preparation, manipulation of the electrode, and the polarity settings.

You will recall that the last digit of the electrode identification code indicates the flux type and welding polarity. It is also a measure of the degree of penetration with a particular electrode because the characteristics of the fluxes affect an electrode's ability to give deep penetration.

When the fourth digit is 0 or 1 the penetration is deep, moderate with number 2, and light with 3 and 4. This ability to penetrate is important. If you were repairing a crack in an automobile's sheet metal body, you would not choose a deeply penetrating electrode for such thin gauge metal; an E6012 or E6013 is commonly used for such work. However, if you were making a frame for a motorcycle carrier, you would need maximum strength to hold the weight of the cycle.

An E6010 or E6011 would give you deeper penetration and added strength.

BUYING ELECTRODES AND STORAGE

Most electrodes are packaged and sold in 50-lb boxes. Some stores package smaller groups of a dozen electrodes or sell them individually. However, it is always cheaper to buy in quantity.

How you store your electrodes is important. *They must be kept dry*. This is especially important if you are storing low-hydrogen electrodes used for high-strength welds. They cannot tolerate moisture from the air and are stored in *drying ovens* after their sealed packages are opened. The ovens maintain a temperature between 280° and 320°F.

WHEN TO DISCARD ELECTRODES

The consumable electrode is gradually used up as it deposits metal in the crater. Most electrodes are 14 in. long when new, although special orders are available in 18-in. lengths. They are discarded when they burn down to about 2 in. At this point they become so short that the electrode holder blocks the welder's field of vision and he cannot see what is taking place. It is wasteful to discard them too early.

PURPOSE OF FLUX IN ARC WELDING

We use flux in brazing and soldering to clean the metal. In arc welding it has a different role. Fluxes are designed in arc welding to shield and protect the puddle from oxygen in the air during deposit and to aid in the cooling process.

Such gaseous shields were not required for oxy-acetylene welding because the flame consumed the oxygen at the weld puddle as it heated the base and filler metal.

The last number of the electrode identification code lists the chemical composition of the shielding material in the flux that is coated to the bare electrode wire. This coating melts along with the electrode metal,

Figure 16-5.

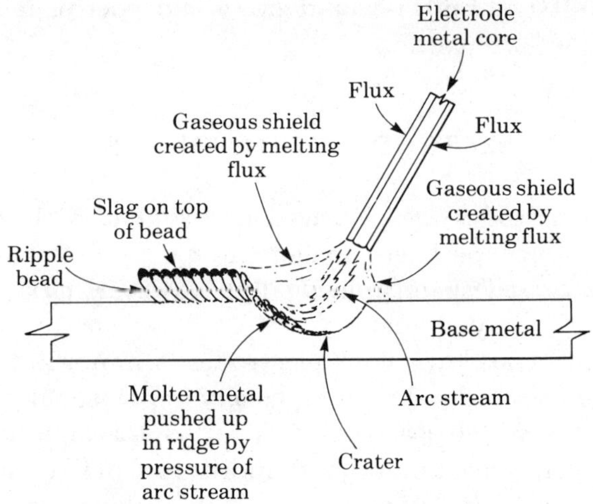

forming an inert gas that excludes oxygen and nitrogen from the air around it.

As the burning of the flux continues, the flux material forms a slag that covers the top of the weld bead. This slag continues to protect the bead from oxygen and nitrogen as the bead cools. In addition, the slower cooling, caused by the insulating blanket of slag, produces a more ductile weld.

REMOVING SLAG

Slag is removed after the weld bead cools. Removal is done with a chipping hammer and wire brush. Lightly tap the weld bead area with the pointed tip of the chipping hammer to break up the slag coating, then wire brush the area to clean away the smaller particles. In the purified state the weld bead can rust easily and should be protected with a coating of paint.

IRON POWDER FLUX COATING

Under the electrode identification code eight types of fluxes were noted. The different types are necessary to take care of special types of welding and techniques.

Among the special fluxes are those containing iron powder. Iron powder flux provides a greater rate of weld melt deposit, which means faster production

Figure 16-6. This chipping hammer is useful for removing any slag deposits that may build up as the weld bead cools.

Figure 16-7. A wire brush makes a very handy cleaning tool for preparing metal before a weld or cleaning a weld joint after a weld.

welding, because a higher amperage rate can be used, resulting in a faster weld bead.

Among the advantages of using iron powder in flux coatings are:

1. Metal is deposited faster.
2. A smoother bead is obtained.
3. It has steady arc characteristics.
4. It requires less clean-up time and effort.
5. There is less arc blow.
6. It is easier to maintain arc length (contact electrode).
7. It can be used with both AC and DC.
8. It is easier to strike an arc.
9. The cooling rate is beneficial.
10. Little if any undercutting results.

The iron powder is added in various percentages to the flux coatings of the electrodes. This means that various electrodes are able to produce different types of weld beads even though they may all contain iron powder (Figure 16-8).

SUMMARY

1. The quality of an arc weld depends upon the use of the correct electrode.

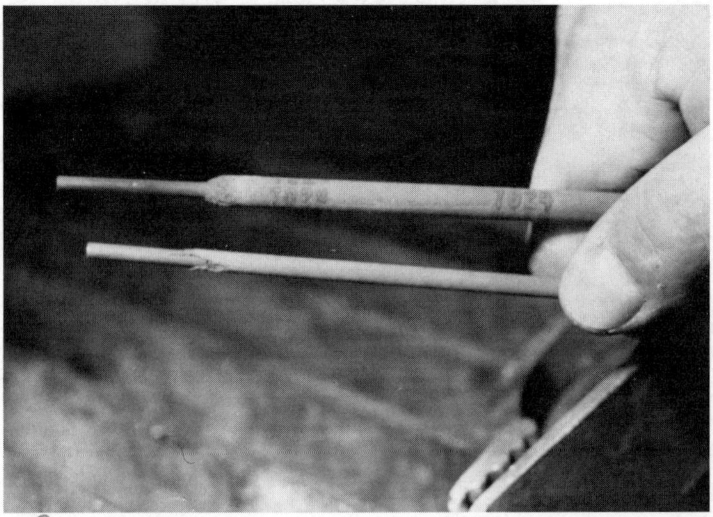

Figure 16-8. Both electrodes shown here are the same diameter, but the iron powder flux coating on the top electrode is much thicker than the standard electrode below.

2. Electrodes are divided into four groups for identification. They are Fast-freeze group (codes E6011, E6010), Fill-freeze group (codes E6012, E6013, E7014), Fast-fill group (codes E6027, E7024), and Low-hydrogen group (Codes E7018, E7028).

3. The code, E6011, for example, identifies characteristics of the electrode. E stands for arc welding electrode, 60 denotes that the electrode has a 60,000 psi tensile strength, the next number—1—shows that the electrode can be used for all positions. The final number indicates the polarity and flux coating.

4. This code is the official designation of the American Welding Society. Manufacturers may have their own designations, but these will include a reference or comparison to the AWS code.

5. Some electrodes may be identified with a color coding.

6. The correct amperage range for various electrodes are shown in the manufacturer's identification and operation charts.

7. Electrode diameter has a decided effect upon amperage. The larger the diameter, the greater amperage passed by the electrode, and therefore, the higher amperage setting required on the welding machine.

8. Excessive amperage causes increased heat, resulting in uncontrollable puddles and increased splatter.

9. Too little amperage does not heat the puddle sufficiently—there is often difficulty in striking an arc and the electrode may freeze to the base metal.

10. Penetration for a strong weld joint depends upon amperage setting, proper joint preparation, electrode manipulation and polarity settings.

11. Flux coatings on electrodes serve a double purpose. Upon burning they generate a gas that shields the weld puddle from oxygen and nitrogen in the air. The flux also forms a slag over the top of the weld bead. This protects the hot bead from the air and slows down the cooling rate to produce a more ductile weld.

12. All electrodes should be stored in a dry place. Lack of moisture is so important for low-hydrogen electrodes that they must be stored in drying ovens.

13. Slag is cleaned from the cooled weld bead with a chipping hammer and a wire brush.

14. Iron powder is added to some electrode fluxes to permit a higher amperage rate that increases the rate of filler deposit and consequently the rate of welding.

NEW WORDS AND TERMS

ASTM American Society for Testing Materials.

AWS American Welding Society.

Electrode The filler metal used in arc welding. It also serves as the current conductor.

Fast Fill An electrode designed to have a large deposit rate.

Fast Freeze Electrode designed to quickly form the bead nugget.

Fill Freeze An electrode in between the fast freeze and the fast fill.

Flux In arc welding a chemical preparation coated on the outside of the electrode to create an inert gas layer to prevent oxygen and nitrogen in the air from weakening the weld. Not to be confused with the flux used for cleaning in braze welding and soldering.

Iron Powder One type of flux additive that provides advantages in controlling weld beads.

Low Hydrogen Another specially designed flux compound that maintains a minimum moisture content.

S Spot The color code spot used to identify electrodes.

Slag A formation of flux and oxygen that deposits a coating over a cooling weld bead.

MACHINE SETUP AND STRIKING THE ARC

Before striking your first electric arc it is important to check your work area or welding station. The area should be protected with curtains to prevent welding light from accidentally interfering with others working in the area. There should be adequate ventilation to carry away fumes and vapors from the flux compounds and molten metal. Check for dampness and moisture. Wear protective clothing, and protect your eyes with a helmet with the proper shade of filter.

Your arc welding work bench should have iron slats or other iron plate surfaces that will stand up to splatter and not catch fire. You will find it helpful to have a few frequently used tools close by, including pliers, vise grips, chipping hammer, and a wire brush.

WELDING MACHINE

Before beginning an actual arc weld you should clearly understand the controls of your welding machine. These will vary with different manufacturers and with the complexity of the machine.

The AC welding machine has the fewest controls. Generally, this type of welder has just an "off" and "on" switch, plus an amperage selector. It is the simplest type of arc welder.

The combination AC and DC machine has the "off" and "on" switch amperage selector and the polarity switch described in Chapter 15. This switch permits the operator to change from AC to DCRP or DCSP.

The DC motor generator arc welding machine may sometimes have a separate control for the voltage in

Figure 17-1. Some machines are wired to produce both AC and DC welding current. These terminals where the cables are attached have covers that have been removed to show how the cables are attached.

addition to the amperage. This is handy when the welder is working out of position and wants to control the molten puddle. By reducing voltage, he can make the weld puddle less fluid and reduce the chance of dripping.

Specialized Machines

There are other arc welding machines for specialized purposes and processes. These processes are beyond the scope of this book. However, some of them also have a conventional "stick" electrode welding capability. If called upon to use them for "stick" electrode welding, you will have to consult the manufacturer's instruction book for operating directions for the arc welding process alone.

Providing Ground

In setting up the machine, the ground cable must be clamped to the welding object (work) or to the table if it is of steel. This will provide the complete electrical circuit from machine through work and electrode and back to the machine.

The next step is to choose the proper type of

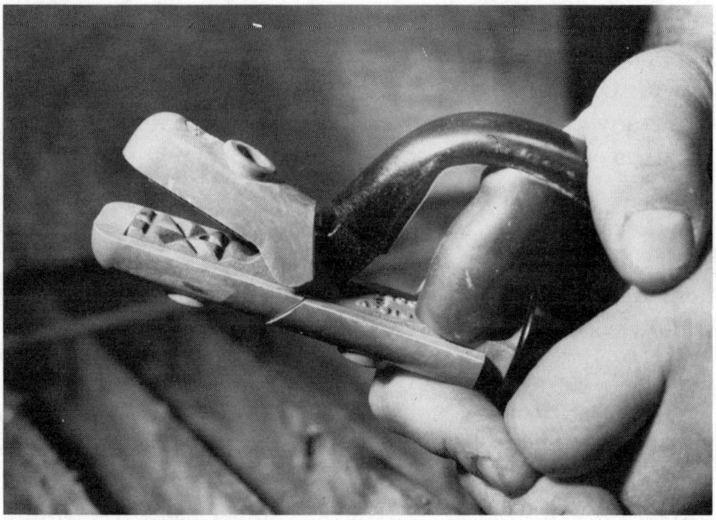

Figure 17-2. The jaws of the electrode holder have several angled grooves for use in correctly positioning the electrode for easy welding.

electrode, according to the identification code, to match the work being done. The electrode must be of the proper diameter. It is then inserted between the jaws of the electrode holder. These jaws have several grooves that are used for positioning the electrode at different angles to work easily in a variety of positions (Figure 17-2).

If using DC, set the machine for the proper polarity if necessary.

It will be helpful to you if you begin with a general-purpose E6011 (1/8-in. diameter) electrode, using 3/8-in. mild steel for the work piece. No joint will be necessary at this time. Begin by practicing laying a bead just as you did with oxyacetylene. For this electrode and metal, set the machine on the AC setting and at 95 amps. This will be adequate.

Handle Electrode Carefully

When you turn on the machine, be careful not to make contact between any part of the table or work and the electrode. If you do, you will form an arc. The flash will tell you that you have made a mistake. If your helmet happens to be up when you make this accidental contact, you will be temporarily blinded. If this happens, close your eyes at once and stand still. Some people become disoriented with their eyes closed. If this happens to you, open your eyes just long enough to sit down. Of course, you must keep the electrode away from any other grounds while this is happening. After about a minute your eyes should begin to return to normal.

It cannot be overemphasized that the light and radiation of an arc are dangerous. They can do permanent damage to your eyes. Every care must be taken to avoid exposure.

STRIKING THE ARC

"Striking the arc" is the term welders use for forming the arc. There are two ways to do this. In one you actually follow a motion like striking a match; that is, you arc the electrode down and scratch it against the metal and then lift it up as the arc forms from the contact. In the second method, you hold the electrode at a 90° angle to the metal and tap it against the work piece to start the arc. The arc is then raised to the proper arc length, which is about the diameter of the electrode.

The proper way to strike an arc is as follows:

1. There is no joint to follow in this first practice, so draw a soapstone line to guide you. Position the electrode at a right angle to the line.
2. Next, tilt the electrode about 20° in the direction of the weld travel. This is approximately the angle you will maintain as you carry the arc forward to make the bead.
3. Swing the electrode down in an arc, scratching it on the metal. The arc *should* appear when the electrode touches the work metal. The electrode must be lifted at once to prevent it sticking to the work piece.
4. Raise the electrode to get the correct arc length of a distance approximately equal to the diameter of the electrode.

Whether you scratch or tap to start the arc is a matter of preference. Regardless of the method, you should practice it until you can form an arc without difficulty.

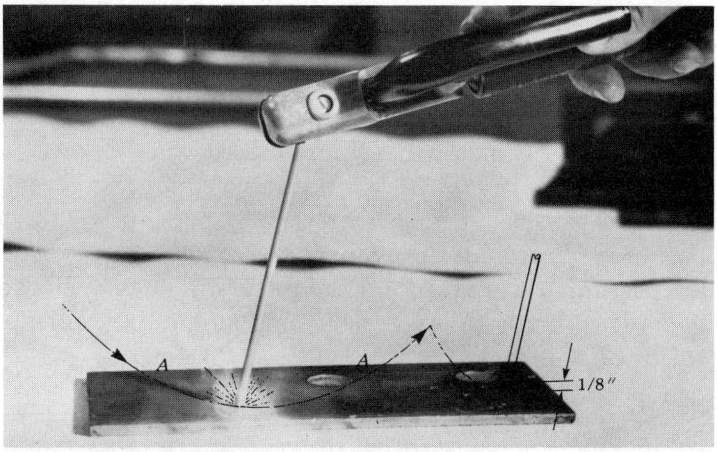

Figure 17-3. In striking an arc, the electrode is swung as you would when striking a match. The arc is created when the electrode scratches the base metal. The electrode is lifted up in the swing and lowered to the correct arc length, usually the diameter of the electrode, but not to exceed 1/8 in.

Figure 17-4. In the tapping method, the electrode is brought straight down against the base metal and then raised as the arc is created.

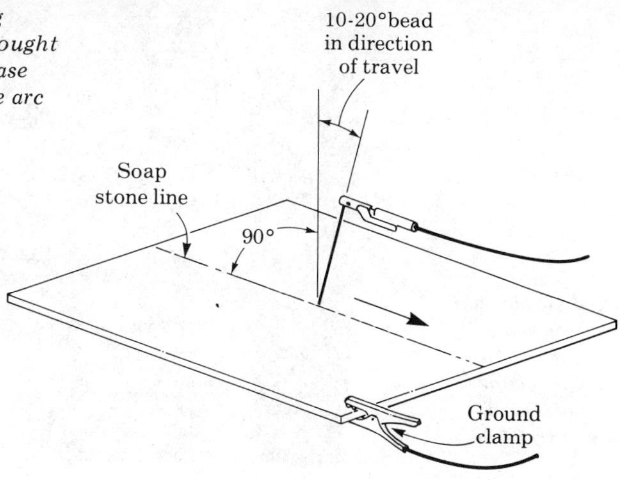

Figure 17-5.

When the Arc Will Not Form

You may scratch or tap and have nothing happen. This generally means that you are not getting any current through your cable or that your amperage is too low.

A break in a cable or a loose cable connection could be the trouble. Or—as sometimes happens with a beginner—you failed to connect the ground clamp. Too low an amperage setting for the metal thickness will also make it difficult to strike and maintain an arc, and the electrode will sometimes flux over at the end.

RUNNING AN ARC WELD BEAD

After you have learned to strike an arc you are ready to practice running an arc weld bead.

You first strike an arc and adjust the air gap to the proper arc length. Heat from the electrode will immediately begin to form a bright red weld puddle and then the arc pressure will dig the crater. The crater is filled with the melting electrode as the electrode is manipulated in the forward direction. The force of the arc stream piles up the molten base metal and melted filler from the electrode in the back of the crater. This piled-up metal hardens into ridges as you move forward with your welding.

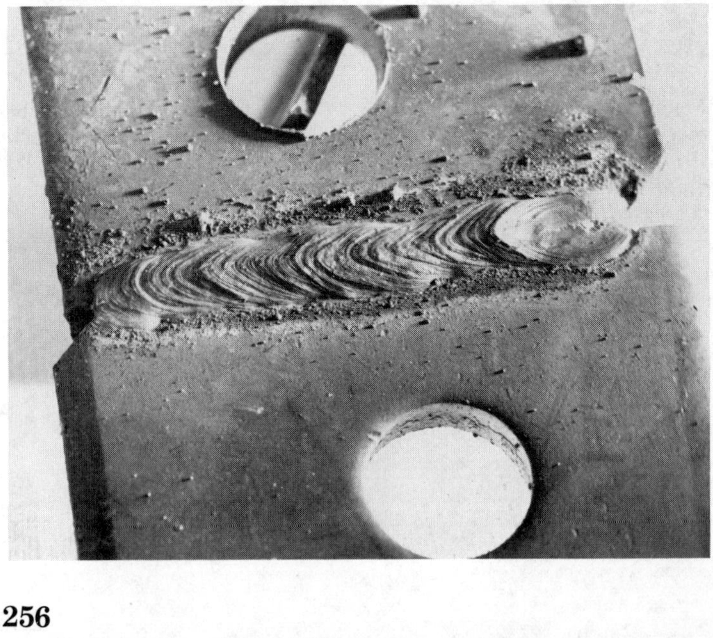

Figure 17-6. Pointed ripples like these are caused by too fast a forward speed. The crater at the right end was, however, properly filled by dipping the electrode in the puddle as the current was cut off.

You will have three elements to control: (1) the angle of the electrode, (2) the air gap, and (3) the forward movement speed.

If you move too fast, the weld puddle will form pointed ripples on the overlapping welds (Figure 17-6). If you go too slowly, the weld puddle will be too wide or may burn through the metal. The normal width of the bead is 2 to 2 1/2 times the diameter of the welding rod. In this particular practice you are using an E6011 electrode with a 1/8 in. diameter. The weld bead width should be from 1/4 to 3/8 in.

After running a bead, you can judge your work for roundness of the overlapping ripples and for weld bead width and height (Figure 17-7).

Undercutting

Undercutting may also appear on the bead margins if the amperage is too high, the electrode is held at the wrong angle, or the weld rate is too slow. Undercutting is the removal of metal from the bead margins of welds as the weld deposit is laid down. Correcting the above-mentioned causes and using correct electrode manipulation will eliminate this trouble (Figures 17-8 and 17-9).

Arc Blow

Arc blow, misdirection of the arc stream by electromagnetism forming in the work metal, has been previously discussed. You will encounter this difficulty only with DC current. If it occurs, follow the procedures listed on page 232. Sometimes the trouble can be minimized by welding in short bursts.

WELD PATTERNS

In your initial practice you will run the electrode in a straight line with no side movements. This is fine to get the feel of laying a bead with an arc, but later you should learn pattern weaving, which will permit you to properly mix your weld materials to form a stronger weld.

Mixing weld puddle materials is accomplished

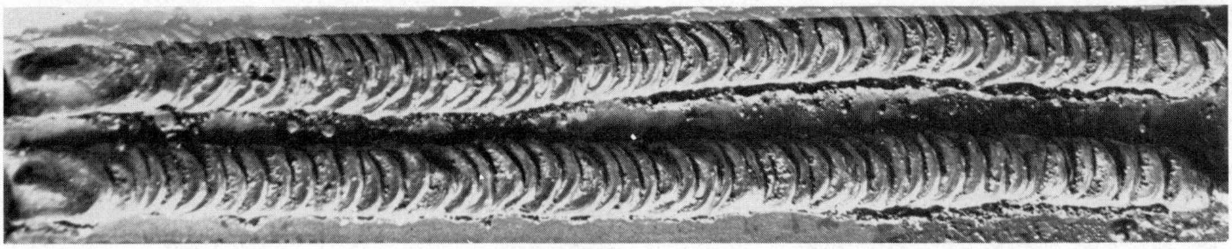

Figure 17-7. This weld bead shows evenly spaced ripples and good bead margin width. This indicates proper forward speed and good arc length.

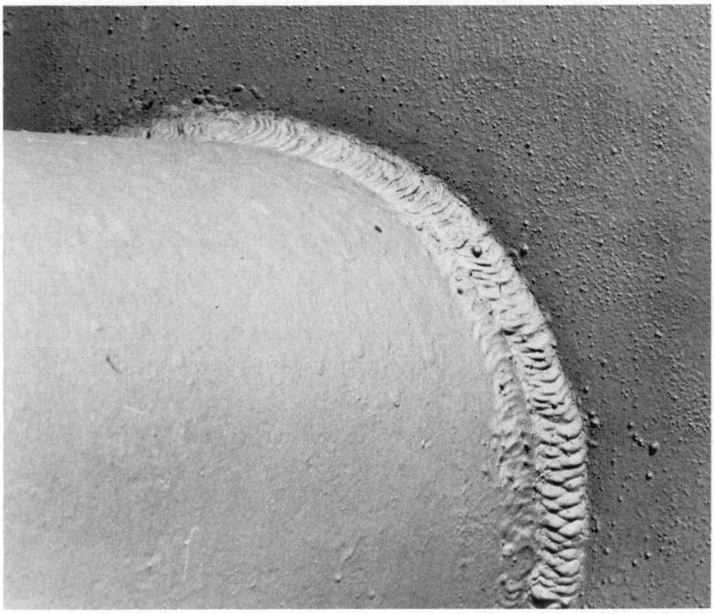

Figure 17-8. In this pipe-to-plate weld, the welder has used multiple bead fillet welds because of the large size of the pipe.

Figure 17-9. Too high an amperage setting has resulted in undercutting, shown as the dark cavity above this fillet bead.

Figure 17-10. *Weave patterns.*

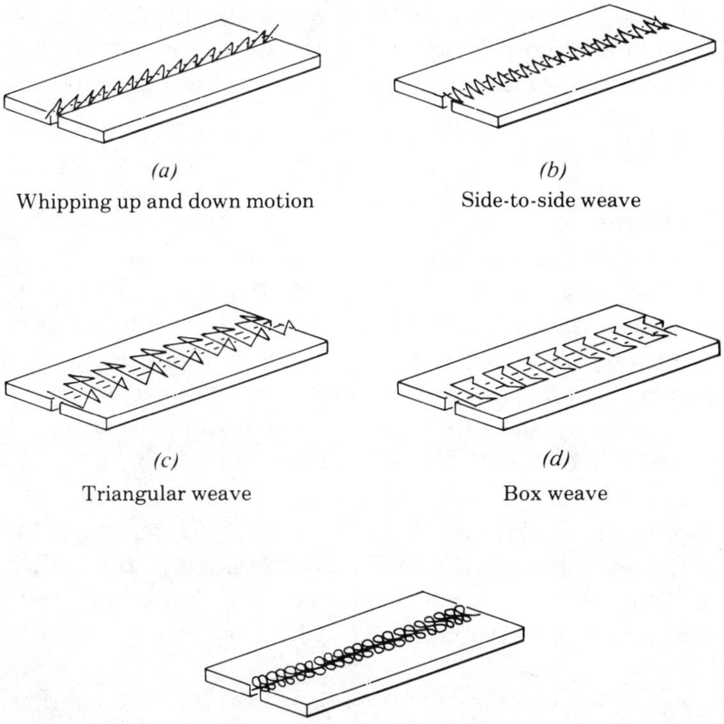

(a) Whipping up and down motion

(b) Side-to-side weave

(c) Triangular weave

(d) Box weave

(e) Overlapping figure 8

through weld bead manipulation. This is done in several ways to achieve a more complete or homogeneous bond. What we are trying to do is not just to lay the filler material (melted electrode) on top of the molten base metal but to mix and unite the two. The result will be a more reliable, ductile weld joint.

The proper manipulation technique is largely a matter of personal preference. When you find the one that works best for you, that is it.

Figure 17-10 shows some of the accepted weave patterns used in controlling the weld puddle through electrode motion: (*a*) is a "whipping" up and down motion; (*b*) is a side to side weave; (*c*) is a triangular weave; (*d*) is a box weave; and (*e*) is an overlapping figure eight.

It should be noted that some electrodes (for example, E7024) are not suitable for weaving manipulation; therefore, the welder should consult the manufacturer's chart to verify any variations in the normal applications of electrodes.

All these weave patterns are made by moving the electrode from side to side as you move forward, except for (*a*), the "whip" technique. In this manipulation the electrode is moved up and down.

Whipping

The slight raising of the electrode (too much will break the arc, of course) allows the weld puddle to temporarily cool between deposits of melted electrode. This is very useful in certain weld positions. It will help minimize sag and drip on vertical and overhead welds.

Not all electrodes are suitable for whip manipulation. Check your AWS identification code on the electrode. If the last digit is 0 or 1, you can whip the electrode successfully. If the last number of the code is 2, 3, or 4, do not whip the electrode. Whipping is not recommended with numbers 5, 6, 7, and 8.

Practice Beads and Joints

You should practice running your beads on the flat surface of the weld metal, using all the various weave patterns. This practice should continue until you can weld a straight bead with even width and height. After each bead is run, you should check and analyze it carefully for all factors, including forward rate of speed.

Figure 17-11. You are now ready to run practice beads. Fill the entire plate with them until you have learned to control the three elements.

Once you have mastered the flat surface weld you should start practicing joint welds on the same thickness of metal. Here you follow the same preweld procedures you did in making oxyacetylene welds. The metal must be cleaned. If it exceeds 1/8 in. in thickness, it must be beveled to secure good penetration. To briefly review, beveling can be done with a grinder or the flame cutting torch. The bottom one-third should be left unbeveled to begin your root pass. The root pass is the first pass of a multiple pass series and forms the foundation for all the others.

In making multiple passes, it is necessary to clean slag and debris from each weld pass before forming a second layer.

Restarting a Weld Bead

There may be times when it is necessary to restart a weld bead during a joint. Restart the arc at the leading (front) edge of the crater, move the new weld to the trailing edge of the puddle, then start making a bead again. This is the safest method to prevent weakening the weld bead by an incomplete bond at the restarting point.

Stopping the Weld

There is also a special technique for stopping a weld. You will leave a crater if you just draw the electrode away. Minimize the crater by hesitating just a moment when the weld bead ends, quickly dip the electrode into the crater, and quickly draw it all the way out. This extinguishes the arc with a minimum of metal transfer from the crater.

BUTT WELDS

In making a butt weld your objective is to achieve the correct width, height, straight bead margins, and 100 percent penetration. If penetration is not complete, check the gap spacing between the two pieces of metal you are welding. You may find it necessary, as in oxyacetylene welding, to tack weld the two pieces to keep the gap constant during the welding pass.

In addition to the proper gap, other factors affecting full penetration are electrode angle (20° in direction of travel), amperage setting, electrode diameter, and forward rate of travel. Full penetration means filling the entire vee groove and the base root to the bottom.

For the butt weld we use the familiar flat bead.

T JOINT

The T joint requires a fillet weld, which you used with oxyacetylene welding, but the electrode is run in a somewhat different manner than for a flat bead. This is necessary because of the increased thickness that must be welded in a T joint. The two pieces of metal to be joined meet in a right angle. The fillet weld is laid along the angle where they join. The puddle is started and extended a distance of about 1/4 in., then the electrode is brought back against the puddle and moved a short distance up the surface of the adjoining metal. Now move back against the main weld puddle and up the wall again. This variation of backhand welding helps to minimize undercutting the wall of the vertical section.

The lap weld, which also employs the fillet weld, is easily accomplished once you have mastered the technique of the T joint (Figure 17-12).

SUMMARY

1. The first requirement is a safe working environment and proper protective clothing.

2. Before beginning to weld, familiarize yourself with the arc welding machine you will be using. Controls differ with manufacturers. The manufacturer's manual will give instruction for properly setting up the machine.

3. The AC welding machine is the simplest, with just an on–off switch and an amperage control.

4. The combination AC–DC machine will have an on—off switch, amperage selector, and a polarity switch. The polarity switch has three positions for AC, DCRP, and DCSP.

Figure 17-12. This is a beautiful example of a fillet weld on a lap joint as well as a pipe-to-plate joint.

5. The arc welding machine will work only if there is a complete circuit. The ground cable clamp must be securely attached to the work.

6. There are two standard methods of striking an arc. One is a motion like striking a match, the other is to tap the arc against the base metal.

7. After the arc is formed the electrode is raised to a 1/8-in. height. This is known as the arc length.

8. The electrode is tilted 20° in the direction of travel. The arc heat melts a puddle and arc stream pressure digs a crater in the molten puddle.

9. The metal piled up by the arc stream solidifies into a ripple bead.

10. Points to watch for are full penetration, proper arc length, correct electrode diameter, correct amperage, angle, polarity, forward speed, and joint gap.

11. The same joints and beads are used in arc welding as in the oxyacetylene process.

12. Restarting an arc in the middle of a bead is done by beginning at the leading edge of the crater, moving back into the crater, and then continuing forward.

13. To extinguish the arc without leaving a large crater, dip the electrode into the crater and then draw it away. This will melt extra metal in the puddle.

WELD INSPECTION

One of the most difficult jobs in welding is determining the quality of a weld bead. A lot can be told from inspection of the surface, but a good surface appearance is no guarantee of a solid weld.

There are a number of ways to test welds, but they can all be grouped into two main categories: destructive and nondestructive testing. Destructive testing means tearing the weld apart to see what is inside. This is fine for spot-checking welding procedures to ensure that you are doing things right, but destructive testing destroys the weld.

Nondestructive testing employs instruments and visual inspections to test the soundness of a weldment but does not harm the weldment. Unfortunately, some of the equipment necessary for nondestructive weld inspection is expensive and may not be available except from the larger industries.

VISUAL INSPECTION

For a beginner, the first type of testing to learn is the visual inspection. As your experience grows, your ability to find flaws in your work increases. Your naked eye can spot undercutting, slag embedded in the bead, blowholes, and cracks; however, you can do much better with an optical aid. A magnifying glass helps (Figure 18-1), but NASA, in the Space Agency's recommendations to contractors working for the space program, advises the use of microscopes with at least a 30 power magnification.

Figure 18-1. A magnifying glass is helpful for detecting minute cracks and imperfections during visual inspection of your weld beads. This is a corner joint.

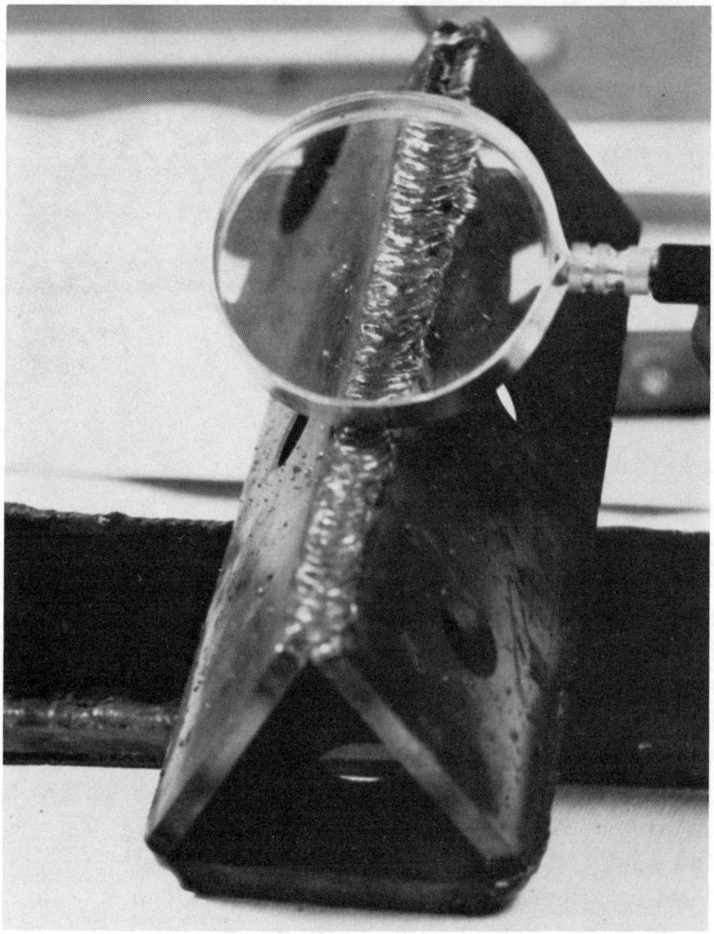

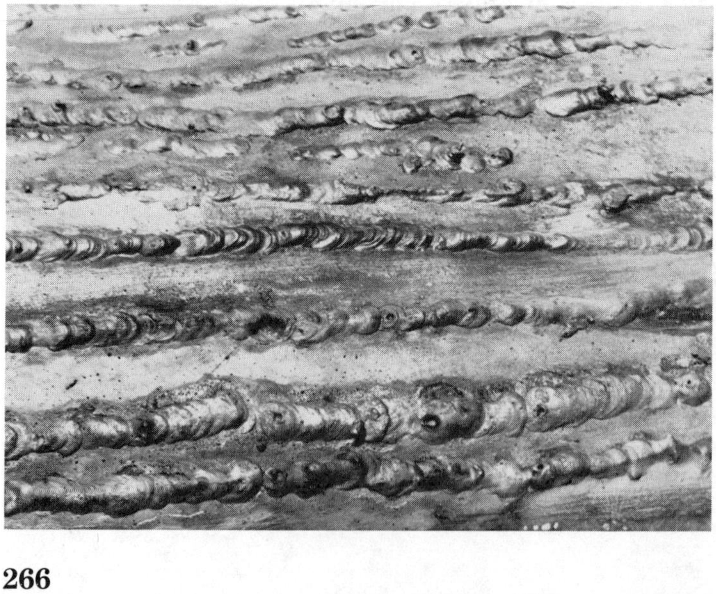

Figure 18-2. These weld bead samples show the effect of unsteady oxyacetylene torch manipulation. The bead width and height is not even. The continuation of the weld bead is broken by difficulty in adding filler rod. The cold weld beads do not blend smoothly at the bead margins, and the hot weld beads are flat without good height.

This degree of blowup will magnify errors that your unaided eye could never see. Tiny hairline cracks or very small blowholes can lead to failure just as quickly as larger defects.

Bead Quality

In making your visual inspection, the first thing to look for is the quality of your bead. The width of the bead and its height should be uniform (this applies to both arc and oxyacetylene weldments). The ripple loops of the weld puddle should be even and regular with no gaps and breaks. The regular width of the weld bead should be about twice the diameter of the welding rod or electrode. The regular height of the bead should be slightly above the surface of the joint and evenly convex.

Outward Indications of Good Interior Welds

If your surface inspection shows good width and height and even ripples through the length of the bead, then this can be used as an indication that your manipulation technique is good; therefore, the weld puddle was properly mixed. If the correct angle and heat were used in the weld, you should expect that you also achieved proper penetration. This can be checked on some welds simply by turning the piece over and inspecting the underside of the joint. Butt welds can be checked in this manner; fillet welds cannot.

Penetration

Penetration must be complete; that is, the weld fusion must go all the way to the bottom of the joint. Penetration depends upon proper joint preparation—gapping and beveling, weld rate, amount of heat, and, in arc welding, the angle of the electrode.

The gap—separation between the pieces being welded—must be sufficient to allow the molten metal to flow all the way to the bottom of the joint. This gap is normally the width of the electrode. However, warpage and distortion can lead to gap trouble, which may cause the welder to alter this setting according to the joint needs.

Figure 18-3. This weld bead shows a very even ripple pattern and bead width, but the joint gap was not correct. This caused the joint to close and distort the gap even further. Notice also the large crater left at the end of the bead.

If incorrect gap was the cause of poor penetration, you can see this when the metal is inspected from the underside. If the unfilled joint is smaller than the width of the electrode or filler rod used, then you may suspect gap trouble as the cause of the incomplete penetration.

Other Causes of Poor Penetration. If the gap is correct and there is still incomplete penetration, then incorrect heat may be the reason. The puddle may not have been sufficiently fluid, which could have been caused by using the wrong tip in oxyacetylene welding or too small an electrode in arc welding. Too-rapid forward speed can result in insufficient melting of base metal and inadequate deposit of filler material. Failure to bevel the metal before welding may also contribute to poor penetration. In multiple passes penetration depends upon correct deposit of the first pass.

Even if the penetration is to the full depth of the joint, the weld is inadequate if the gap at the bottom shows any cracks or holes.

Craters

A crater is the depression left at the end of a weld bead. It is caused by the gas pressure or the arc stream that pushes back the molten puddle to create the characteristic ripple bead. When the heat is stopped to end a weld, the crater freezes in its depressed state.

A crater will always be present, but you should do what you can to minimize it. In oxyacetylene welding

Figure 18-4. The weld bead has good width and height, but the crater shows insufficient metal. This crater has the beginnings of a crack forming in the center.

extra filler rod can be melted into the crater when the bead is ended. The same effect may be obtained in arc welding by dipping the electrode into the puddle and quickly removing it before it freezes into the cooling metal. Either of these techniques will minimize the amount of depression left at the end of the weld bead.

Blowholes

Blowholes occur during bead formation. They are the result of excessive gas pockets trapped in the molten metal as it begins to solidify. The trapped gases expand, creating pressure that bubbles its way upward. The bubbles break through the surface, leaving a "blowhole" (Figure 18-5).

A blowhole weakens the weld zone for the bead no longer has a continuous bond. The joint then becomes like a chain that is no stronger than its weakest link. Blowholes occur more often with certain types of alloys which produce excessive gas formation. Controlling your welding rate so that you heat long enough to cause any gas bubbles to rise to the surface and burst before the metal solidifies is one way the problem can be minimized.

Sometimes the trapped gas does not get to the surface. When the bubbles are trapped below the surface, it is called an *inclusion.* These inclusions also weaken the weld joint.

Weld with the proper weave pattern or manipulation

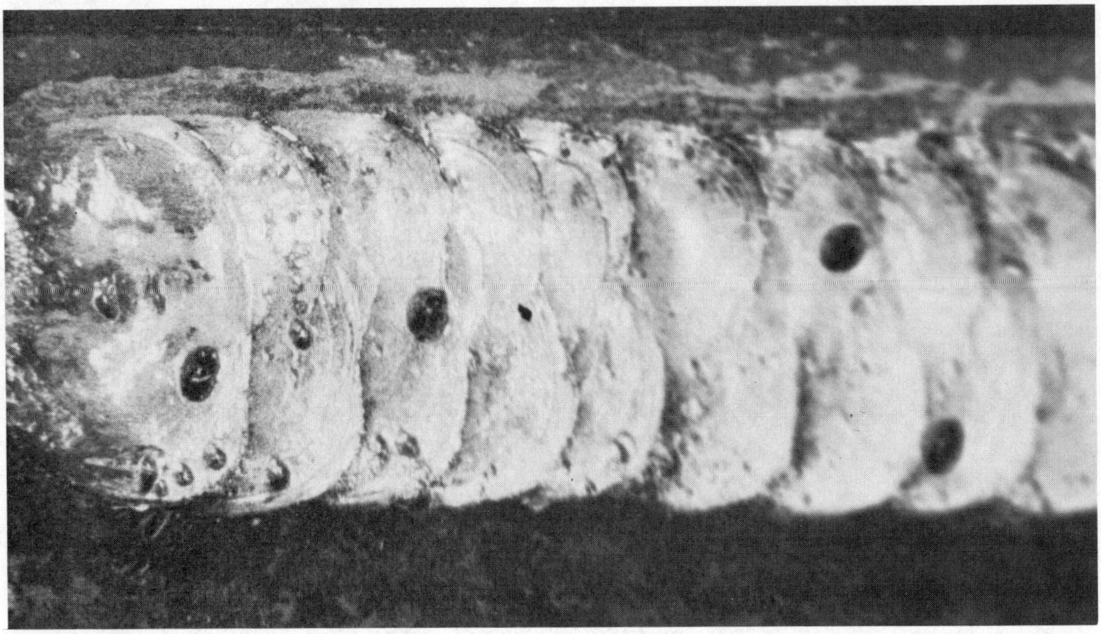

Figure 18-5. This is a close up of a blowhole in an arc weld. The small bubbles left on the surface are the exit for the trapped gas.

of the electrode. This manipulation mixes the molten weld puddle, assuring movement of heat transfer, and will enable the gas that forms to rise and escape through the surface.

Undercutting

Undercutting is when the flame or arc eats out a groove from the actual weld bead margin. Figure 18-6 shows undercutting on a fillet T joint weld and a butt weld.

Undercutting may be caused by several things. In oxyacetylene welding on a fillet bead, sagging of the upper portion of the puddle exposes a melted portion of the vertical metal. A tip that is too large may also do the same thing. In arc welding undercutting may be caused by excessive amperage or improper electrode angle.

Undercutting is not to be confused with overlapping, which is caused by current set too low. If your visual inspection shows that the actual base metal at the margins of the bead has been melted away, that is *undercutting.* But if the base metal is not cut and the undercutting extends under the bead itself, then this is *overlapping.* Figure 18-7 shows undercutting, overlapping, and a sound weld.

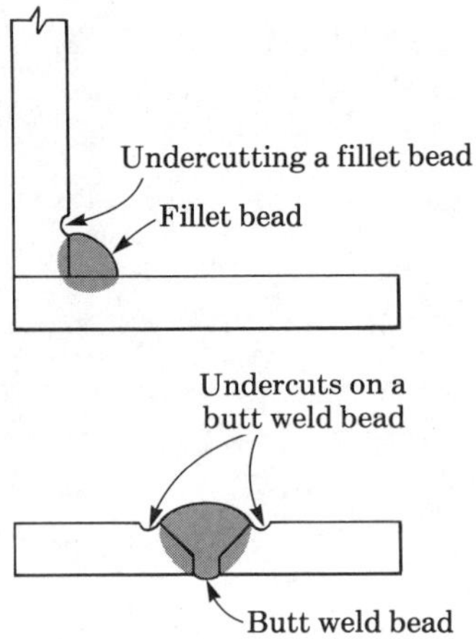

Figure 18-6.

Some Bead Errors

Figure 18-8 shows some of the mistakes a beginner makes. One of the most noticeable is that the beads do not run straight. This can be easily corrected by drawing a soapstone line for a guide.

Another thing that stands out in this photograph is the flat-spreading beads without ripples. Here the trouble could have been caused by too much heat which produced an uncontrollable puddle that spread instead of piled up. Unless special requirements dictate, the normal width of a weld bead is about 2 to 2 1/2 times the diameter of the filler rod.

There are also numerous gaps where the bead has been broken. Here the torch or electrode was probably lifted too high.

The blowholes and large end craters have already been discussed.

Figure 18-7.

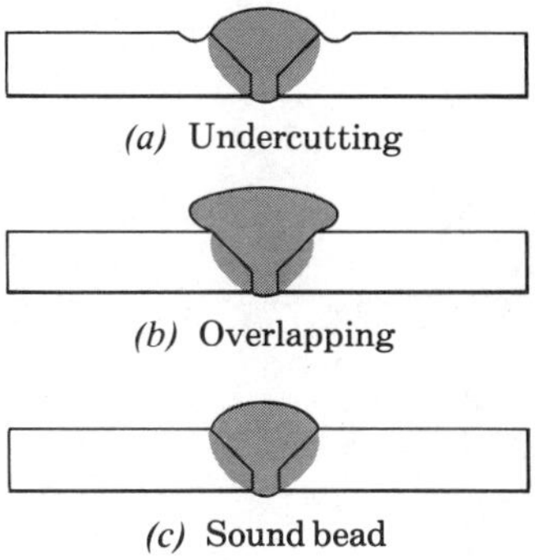

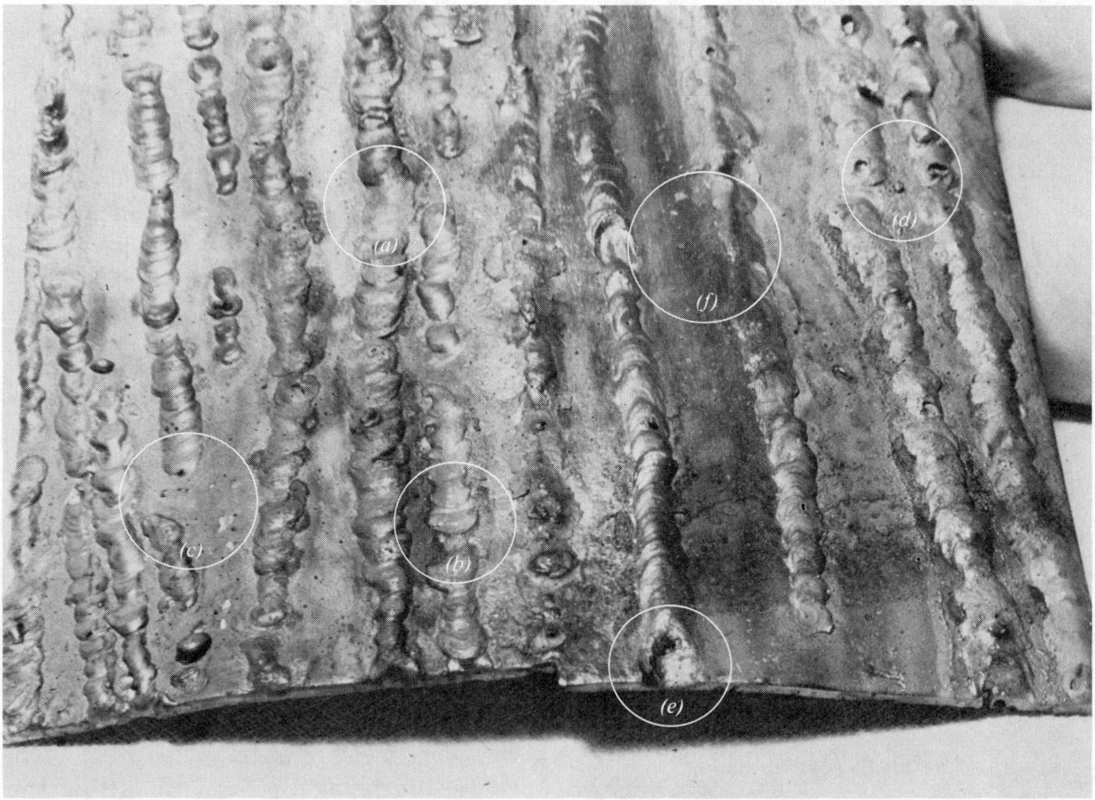

Figure 18-8. Some common errors of beginners are shown here. They include crooked lines (a), spreading beads without ripples (b), gaps (c), blowholes (d), large end craters (e), and improper filler manipulation (f).

OTHER NONDESTRUCTIVE TESTS

These visual tests are ones that every welder can perform on his work. There are other tests that will reveal defects which cannot be detected through visual inspection; unfortunately most of these require specialized equipment that will not be available in small shops. Some are so expensive and bulky that you will find them only in very large industrial plants. Among the tests are x-ray inspection, ultrasonic inspection, magnetic particle inspection, and liquid penetrant inspection.

X-ray, ultrasonic, and magnetic particle inspections can show defects under the weld and inside the bead. Liquid penetrant shows minute cracks and surface defects invisible to the unaided eye. A special dye substance is painted over the weld. Capillary attraction draws this penetrant into the defects. A developing substance added later makes the penetrant visible in the crack. A variation of the system uses ultraviolet (black) light to make the penetrant in the cracks flourescence.

Macroinspection

A cross section of the metal is etched with nitric acid to bring out the grain structure. This section is then inspected under a low-power microscope or magnifying glass from 5 to 30 power. You can see the penetration and fusion of the weld metal to the parent metal. Slag or gas pockets become visible also. The disadvantage of the process is that you can study only small prepared segments.

Other Defects

There are a number of other defects to be found in improper weldments. Some of these are peculiar to certain processes; for example, the *offcenter weld* is found in welds made by some automatic machines that use double electrodes. It is caused by the material not being centered between the electrodes.

Another defect found in arc welding and oxyacetylene welding is the *open weld.* This happens when there is a misfire or a circuit break or even by lifting the electrode too high and breaking the arc. In some cases it is just an open gap where no fusion takes place. In others (Figure 18-9), as in this oxyacetylene bead, a backfire in the torch has left a crater and a weakened, distorted section of bead.

Metal expulsion is where excessive pressure causes excessive bulging. Globs of molten metal are pushed from the crater. It is not to be confused with bulging of base metal caused by distortion, where the trouble is warpage due to uneven pressure. Metal expulsion on flat beads may look like the effect of metal sag on vertical welds. *Splattering* is a milder form of metal expulsion.

DESTRUCTIVE TESTING

Destructive testing, which means destroying the weld joint, is essential to the beginning welder. Visual inspection can reveal many defects, but you must know what is inside so you can correct your errors as you advance.

Destructive testing is impossible on every weld, but it should be done on selected samples even after some degree of proficiency is achieved.

The bend test, which involves placing a welded piece in a vice and bending to a full 90° angle, was described earlier. To review briefly, this bending of the weld joint will disclose any weaknesses. Cracks may form. Within the crack look for signs of blowholes, inclusions, penetration, and grain size. If the grain size is coarse and loosely packed, it indicates that improper cooling took place. This could mean that the weld sample was heated excessively or cooled too rapidly. This test is practical only on low carbon (mild) steel. High carbon steel, with its lack of ductility, cannot be bent in this manner.

After the sample has been bent in one direction to a 90° angle, it is bent back 90° in the other direction. Low carbon steel should be able to take this double bend without cracking if the weld is good. In all cases a good weld should be as strong as its parent metal.

Guided Bend Test

In this test a special bending jig with a hydraulically operated ram bends a piece of 3/8-in. welded low carbon steel. American Welding Society specifications

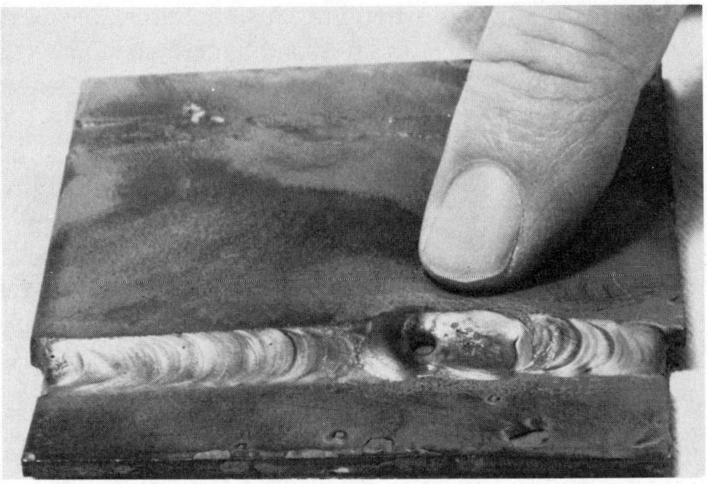

Figure 18-9. This lap weld shows a beautiful fillet bead that is lost because the torch was held too close to the surface and caused a backfire. The backfire left a large crater, which weakened the joint.

Figure 18-10. Splattering is a mild form of metal expulsion.

require that the welded steel be bent into a U-shape with a 3/4-in. bend radius. The metal should bend without forming any cracks or breaks along the weld joint.

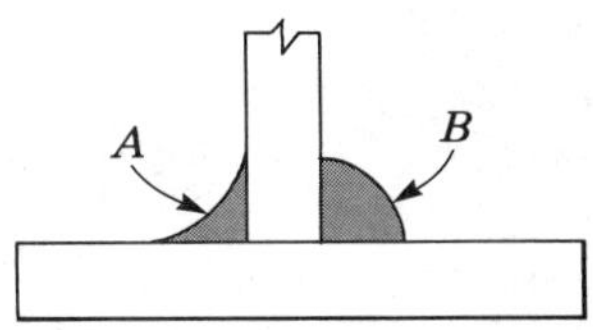

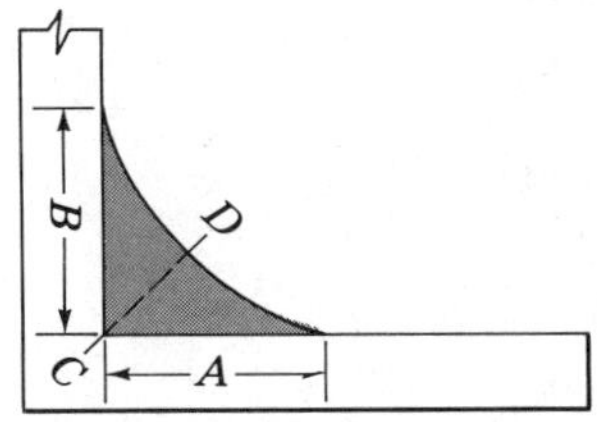

Figure 18-11. (a) A shows the concave fillet bead. B shows a convex fillet bead. (b) In a well-made fillet bead, leg A will be equal in length to leg B. The distance across the middle of the bead, C to D, should be about three-fourths the length of a leg.

Testing Fillet Welds

A standard fillet test begins with a visual observation of the bead shape. The bead may be either concave or convex (Figure 18-11). The bead width should be about equal on each leg and these legs should be about twice the diameter of the electrode. The thickness of the bead in the center between the two margins should be about three-fourths of the length of a leg. Check for undercutting, blowholes and craters.

The bend test for fillet welds can be made as shown in the accompanying drawing. The crosspiece is placed in a vice and a bending bar is used to bend the projecting metal until it is parallel with the clamped piece. If it does not break or crack, it is bent back in the opposite direction, then returned to its original position.

Even though the parts bend without breaking, this test should still be considered *destructive* testing. The bending of the parts will subject the metal to compression, expansion, and stresses that will weaken the metal just as bending a piece of wire back and forth will finally break it in two. After this test the weld and metal should be discarded or used for further welding practice. It should not be put into any kind of work that will be affected if the piece fails in the future.

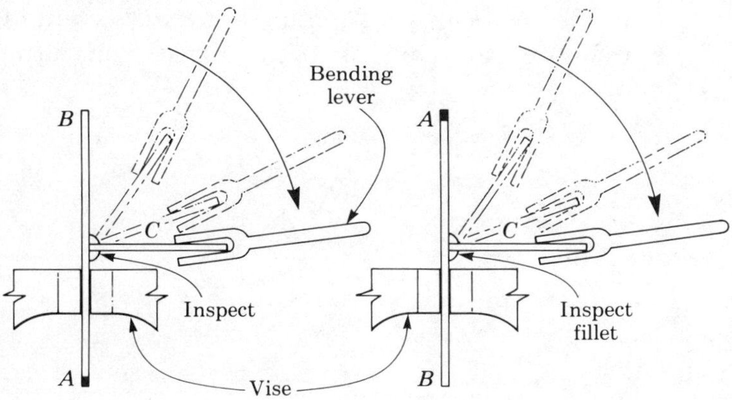

Figure 18-12. Bend test for fillet welds.

SUMMARY

1. Industry has standards for welding quality to give proper service to customers and in construction work that involves welded joints.

2. Although a beginner's work may not fall into these classes, testing to ensure that his work is of industrial quality is essential.

3. Weld testing is grouped into nondestructive testing and destructive testing.

4. Nondestructive testing involves visual checks to find flaws with the naked eye, inspection through magnifying glasses and low power microscopes, x-ray, transonic, and other means. These tests do not harm the weld bead and the inspected item can be used.

5. Destructive testing involves bending, pulling, or breaking a weld joint to determine its penetration, fusion, and freedom from faults such as blowholes, inclusions, cracks, undercutting, and overlapping.

6. Destructive testing ruins a piece for further use. Consequently, it is used for sample testing to indicate that the welder's technique and production is adequate.

NEW WORDS AND TERMS

Blowholes Defects caused by gas bubbles that are trapped in the solidified metal.

Crater The depression left at the end of a weld bead.

Guided Bend Test Bending a welded sample to determine the strength of the weld bead. An example of destructive testing.

Macroinspection Use of a magnifying glass of 5 to 30 power to inspect grain structure in a cross section of a weld sample that has been etched with nitric acid.

IV

SPECIAL APPLICATIONS

ART METAL WELDING

One of the most exciting and certainly the most novel development in the world of art is the booming interest in the welder's torch as an artist's brush. This new art is a development of the last few years, and the results are seen in bric-a-brac stores as well as in great museums. The techniques of welded art are not difficult—they require no more from a welding standpoint than you have already learned.

Art metal work takes many forms. It may be metal sculpture, in which the artist welds great sheets or plates together to achieve colossal effects. It also is putting pieces of junk together to make comic figures. It may be bending and welding strips of wrought iron to make decorative fences, decorations, or even very elaborate gates (Figure 19-1). This huge gate in Taiwan is made of wrought iron, which was then painted selectively in black, gold, and brilliant vermilion.

BASIC ART METAL TECHNIQUES

Every technique you have learned in welding can be adapted to art metal work. For most art work you will use the oxyacetylene torch. Arc welding is too hot for this type of work, although it can be used to an advantage in building wrought iron fences, furniture, or in working with heavy metal plate.

Oxyacetylene is used where detail is required. For smaller designs or where the yellow appearance of gold is desired, we use a brazing technique to take advantage of the yellow-red color. Also, in making comic figures of

Figure 19-1. This Chinese gate is an example of elaborate art metal work. The gate itself is made of an intricate combining of several wrought iron designs. Flat sheet metal was then tack welded to the wrought iron frame. Black, gold, and vermilion paint was used to finish the design.

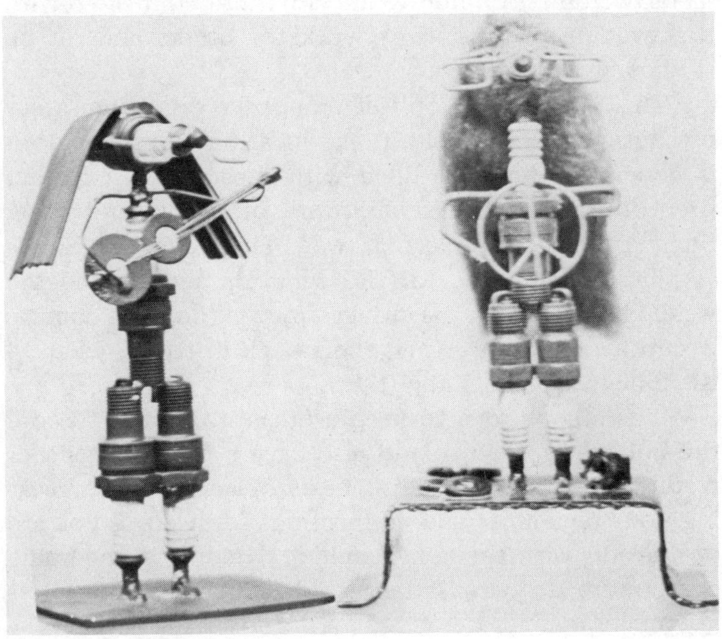

Figure 19-2. Brazing can be used to put junk together to make comic figures.

junk, we may have incompatible metals. In such cases brazing is the answer.

On small figures and objets d'art, soldering may be the most practical method and may be done with either torch or soldering iron.

ART METAL EQUIPMENT

On large object work you may use your regular oxyacetylene torch and tips, but when making small objects, you need a small torch with special tips. This will permit you to work in close on small delicate parts where space is confined. The reduced heat of the small torch is essential.

A cutting torch is also a necessity in working with metals thicker than thin sheet which can be cut with sheet metal shears. A cutting attachment for the welding torch is sufficient and a lot cheaper than buying an entire cutting torch.

A PLACE TO WORK

The welder who is fortunate enough to own his own shop has a place to work, but many who take up art metal welding as a hobby are either not connected with the welding profession or work for others and set up studios in their garages.

This is fine, but only if one observes all the safety precautions. Never weld in any location unless the floor is concrete, the walls lined with sheet metal or other fireproof material, and all trash, oil spills, and other flammable material cleared away. There must be ample ventilation. Curtains for the windows and around the welding position are essential to prevent any eye damage if curious children or neighbors peek in to see what all the flickering light is about.

It is a good idea to keep a bucket of water to cool the hot metal. You should also have a fire extinguisher readily available. You must be as diligent about checking your regulators and connections for leaks as you are on the job with the boss watching. Welding in the home is no more dangerous then welding in the shop unless you make it so.

STICK WELDS

One of the simplest types of art creation is stick figures and objects. These are cartoon lines made by curving and welding filler rods of iron, brass, or bronze. You can also use silver, if you are rich enough.

A stick weld is made by fusing filler rods to form the outline of an object, then flat pieces of metal are added for accents, such as the metal sails on the rodded outline of a sailing ship in Figure 19-3.

Objects like these are easy to make. You can copy designs or you can make your own. A ship is easy to design. Find any drawing or photograph of a boat in profile, lay a piece of tracing paper over the picture, and draw around the outline with a felt tipped pen. Take the resulting drawing (Figure 19-4) and enlarge it to the desired size, which becomes a pattern for measuring and cutting your welding rod to the correct size. Then it is just a matter of welding, brazing, or soldering the rods together.

The sails are cut out of flat metal and spot welded on. In this drawing they appear to have been set at an angle. This is not true. They were cut at an angle to give this appearance. They are flat pieces of metal tack welded to the masts.

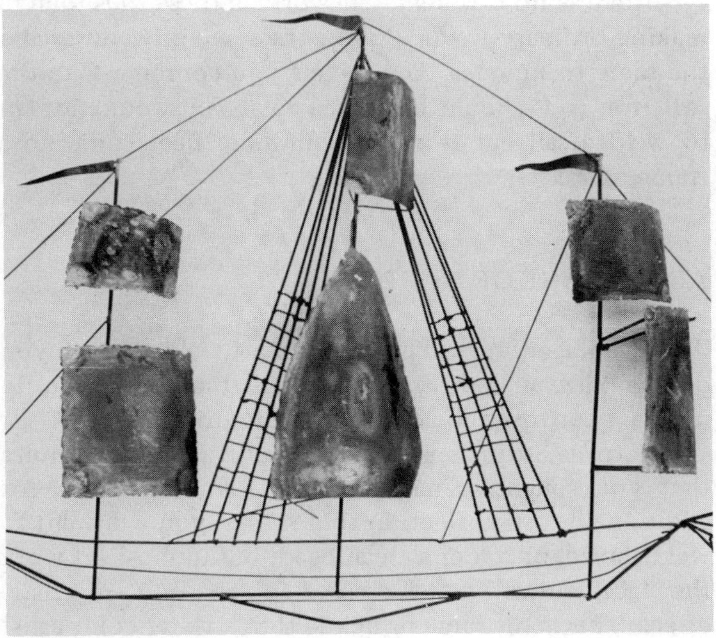

Figure 19-3. Stick cartoon "drawings" made with welding rods are easy to construct. This boat contains only rod and a few flat pieces of sheet metal for sails.

Figure 19-4. This drawing for a stick weld of a ship is a variation of the ship in Figure 19-3. The sails appear three dimensional but actually are flat. They were cut at an angle to give the appearance of dimensions.

WORKING THE RODS

Making objects of this kind is as much bending as it is welding. If you have difficulty bending heavier rod, heat the joint with the torch until it is soft.

Brazing rods with their brassy colors make very good working material, but you can even use coat hangers if nothing else is available. They weld nicely, but they should be cleaned of any paint to get a good joint.

There is no difference in working these rods than in making ordinary welds. Follow the same procedures and the same techniques. The metals you combine with the soft iron rods should be compatible. You would not try to weld a sail cut from an aluminum beer can onto a frame of steel.

ADVANTAGE OF WARPAGE

An advantage you will enjoy as an art welder which you do not have in industrial welding is that you often do not have to worry about warpage and drip.

Warpage, drip, splatter, and burned through holes that you sought so hard to avoid can all be used for additional artistic effect. In some cases, you will want to weld, braze, or solder a clean bead, but in most art work the "accidentals" add increased interest and an artistic effect. This is the same principle that a water color artist uses. He does not lavishly try to make every stroke precise, but weaves accidentals that arise from the flow of color into his compositions. You cannot, of course, be too sloppy.

TIN CAN ART

Just as you can use wire coat hangers in place of welding rod for thin stick and twisted wire figures, you can also use tin cans. These containers are actually steel with a tin alloy coating. Aluminum cans require an aluminum alloy rod for welding.

When working with tin can metal, solder or brazing is better than welding because the metal is very thin and difficult to weld without burning through the can. The tin alloy coating makes a good surface for soldering.

Figure 19-5. Flowers make good projects for tin can metal art.

How to Make a Metal Flower

Flowers are good projects for tin can metal art. Hobby shops have pamphlets that give patterns and instructions for making paper flowers, and these patterns can be used with minor adaptations for tin can art.

A common art metal flower is the rose (Figure 19-6).

Figure 19-6. Flowers are easy to make. This rose was made from the cut tops of tin cans. Brazing rod was used to tack together the lids.

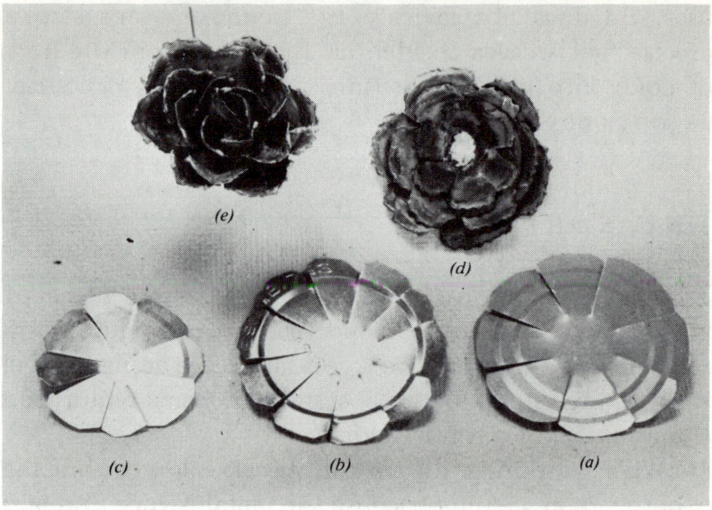

Figure 19-7. The different steps in making a tin can rose. (a) shows a can lid split by tin shears. (b) shows a second lid but smaller and placed on top of the first. (c) shows the shape of a third lid cut still smaller, to be placed on top of the others. (d) shows the three leaves (lids) fused together, with a center brazed spot weld. The edges are touched with the torch to give texture. Then pliers are used to blend up the petals to create the final effect (e).

It is often recommended to beginners because of its simplicity. To make the blossom shown in Figure 19-6, the only tools you need are a can opener to remove the lids from three cans, tin shears to shape the lids, soldering equipment, a pair of long-nose pliers and—if you are brazing instead of soldering—a brazing rod.

Follow this procedure (Figure 19-7):

1. Cut the end from cans. Using the tin shears, make eight cuts toward the center (*a*). This provides the base.
2. Trim the top of each petal in a slight curve (*b*).
3. Trim the second lid in the same manner, but smaller than the others (*c*).
4. Braze the three layers with a tack weld in the center (*d*). The torch will blacken the metal, producing a pleasing dark color.
5. After the tack has hardened, take a pair of pliers and carefully bend each petal upward. Work gently to avoid pulling the tack welds loose.
6. Bend each petal upward and also in a slight curve, as shown in (*d*). The center petals are bent more and closer together to produce the bud effect (*e*).

Although tin cans were used here to illustrate that the cheapest salvage material can be used for art metal projects, you can use any thin metal sheet to make these flowers. Copper makes exceptionally attractive metal work. A worldwide shortage of copper, unfortunately, has caused prices of the metal to increase sharply in recent years.

COMBINING WELDED ART WITH NATURAL MATERIAL

Your welded art can be combined with or mounted on natural material to extend their beauty. Figure 19-8 shows an example of this. A butterfly has been made of sheet metal covered with golden brass and then mounted on a section of tree root. The dark red of the root and the brassy sheen of the butterfly combine to make an effective combination.

Objects of this kind are easy to make. The shape of the butterfly was first traced on paper, the outline drawn on the metal with soapstone, and then the outline was sawn out with a hacksaw or cut with the cutting torch.

Next, the resulting metal outline was tinned with a brazing rod to give extra body and a golden color. The edges were permitted to curl and warp a little in the heat of the torch before the brazing material was touched to the hot metal.

The antennae were small pieces of brazing rod curled around a larger rod as a form. The straight ends were then tacked to the underside of the butterfly between the two wings.

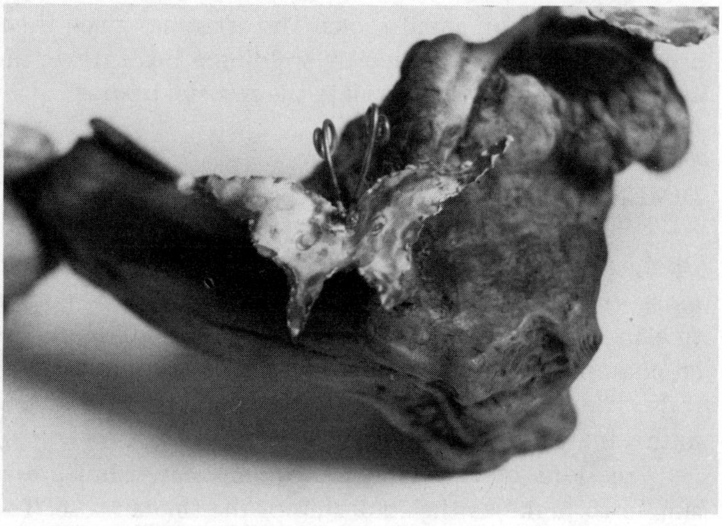

Figure 19-8. This golden butterfly is made of sheet metal tinned with brazing rod. The antennae were curled around a larger rod and then tacked to the wings. The completed insect was then affixed to a tree root.

Figure 19-9. These butterflies are made of tin can metal. The dark metal was highlighted by touching the hot surface with the brazing rod to create golden spots on the wings.

Mounting

The insect was mounted by selecting a suitable gnarled tree root with a color that contrasts with the golden sheen of the butterfly. A small hole was bored in the root. A piece of welding rod was inserted in the root and brazed to the underside of the butterfly.

Welded objects can also be mounted on pieces of driftwood for beautiful effects. A wood screw can be started into the wood and the art metal brazed to the head of the screw.

Figure 19-9 shows two other butterflies that have been mounted on small rocks. The artist has given them a more lifelike appearance by welding a bead down the center of the wings to simulate the insects' bodies.

DIMENSIONAL CREATIONS

Going a step beyond these simple objects, you can build more elaborate, full-dimensional creations. The golden tree in Figure 19-10, seeming to grow out of a rock, is an example. This tree is constructed entirely of brazing rods. The golden color is part of its attraction but is lost in this black and white reproduction.

Creations of this kind look complicated but are as simple to make as the stick ship or the tin can rose. The

Figure 19-10. This golden shower tree is easily constructed of bent braze rods that are thickened by dripping filler. The leaves are splatters of thin filler.

basic trunk of the tree is made by bending welding rods into the desired shape. Then individual pieces of rod are cut, bent, and brazed to the main trunk to provide limbs according to a rough sketch (Figure 19-11). After this basic trunk and branch system have been assembled, the form is modified according to the artist's desires. This may be necessary because outlines drawn on a flat surface often look different when constructed in three dimensions.

Dripping the Trunk

The trunk is built up with the torch and brazing rod to produce a gnarled, lumpy appearance. This is done by

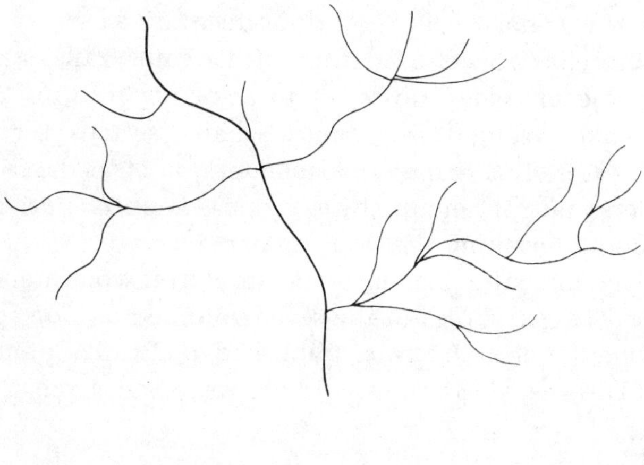

Figure 19-11. Basic pattern for a tree.

overheating the filler and permitting it to run and sag. The rod is also moved and raised and dipped during application to produce lumps and unevenness. The trunk is made thicker at the bottom, gradually diminishing toward the top, where it fades into limbs.

A second large limb is brazed out from the basic trunk, but the remainder are left the size of the original rod, although the rod may be heated, warped, or touched here and there with filler to give it shape. In this particular example, the artist-welder has attempted to break up the evenness of the limb rods by brazing circular lumps along their length between the leaves.

Creating the Leaves

The leaves are made by melting brass rod with the torch and letting the filler drip and splatter on a fire brick. It will not stick to the surface and the brick's texture will be imparted to the splattered filler. Several splatters can be blended together to create as large a leaf as desired. It is not necessary for the leaf to be solid. Open areas where the splatter has not fused may add additional interest (Figure 19-12).

WROUGHT IRON

Wrought iron is pure iron mixed with iron oxide and silicate slag. The slag causes a reaction with the iron to produce a tough, ductile metal that is corrosion resistant. Wrought iron handles with mild (low carbon) steel are easy to weld.

Wrought iron's corrosion resistance, its ability to resist breakage, and its ease of working combine to make this metal ideal for outdoor decorative use.

Wrought iron is used for fences, gates, rails, stairs, wall objects, fancy posts, or just for decoration. You can make wrought iron tables, chairs, settees, scrolls, book cases, flower pots and planters, pots, beds, room dividers, mirror and picture frames, lamp stands, and just about anything else you can name.

Students interested in ornamental iron work are well advised to get copies of the seven-volume *Book of 1600 Ornamental Iron Designs*, published by R. J. Cunningham Designs, 3881 South 3200 West, Salt Lake City,

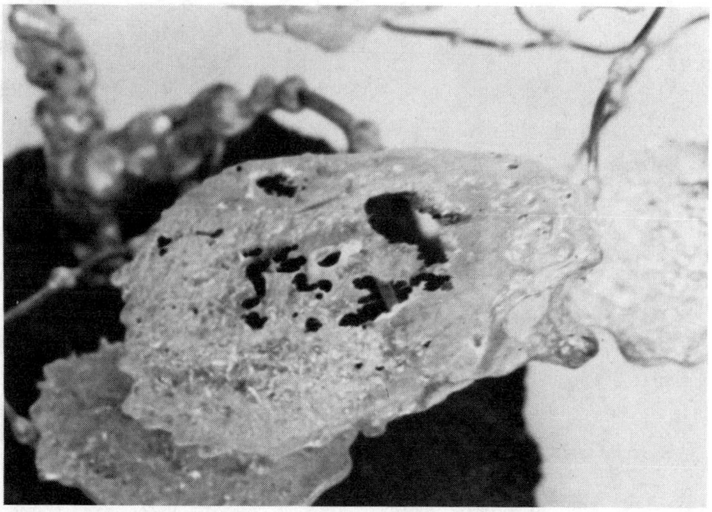

Figure 19-12. Leaves are splatters of molten rod. After the material hardens and cools, it is brazed to the tree limbs. Leaves can be made solid or left with open areas to add interest.

Utah 84119. These books contain pattern designs for nearly every type of wrought iron work.

Working Wrought Iron

Welding wrought iron is similar to welding mild or low carbon steel. You may use either oxyacetylene or arc since you are working with thicker metal in wrought iron.

Fences and ornamental rails and dividers for the outside of the house are usually made of 1-in. square rail, braced together with 1/2-in. square tubing as ribs.

In the accompanying photograph, this wrought iron fence was made by welding 1/2-in. square tubing to brace rails at top and bottom. The flat arrows, giving design to the top, were cut from 1/8-in. flat stock and arc welded to the top rails.

The fence was made in sections in the welding shop. The individual sections were spot welded together (Figure 19-14) in the field with a portable arc welding outfit.

Figure 19-13. Wrought iron lends itself well to making flowerpots and planters.

Cold Working Wrought Iron

Because wrought iron has good ductility, it can be bent without breaking. This permits us to form pieces for

Figure 19-14. In making wrought iron fences, the fence is welded in sections in the welding shop; then the sections are welded together in the field. Note the C clamp used to hold the sections rigidly in place until the weld is completed.

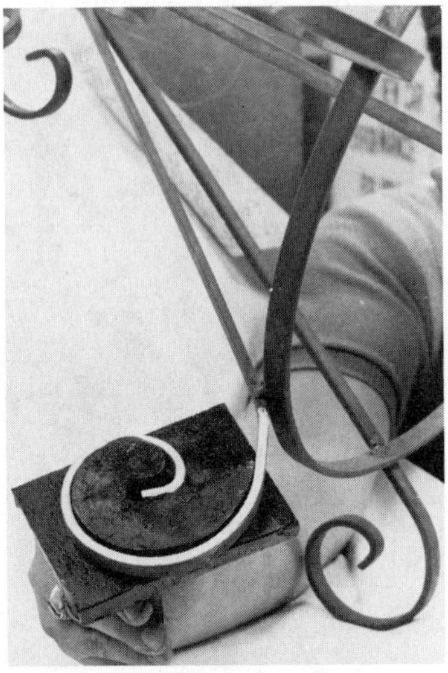

Figure 19-15. The leg of a wrought iron table is being formed around a jig. If the metal is too thick for cold bending around the jig, the wrought iron may be heated until it is soft enough to bend.

scrolls, table legs, and decorative effects without heating the metal. Relatively thin pieces can be shaped cold by bending them around jigs and forms (Figure 19-15). This demonstration shows the welder holding the jig. In actual practice, he would secure the jig form in a vise and then bend the iron around it. The article being cold worked here is the base of a wrought iron table leg.

If the metal is too thick to cold work, it can be heated with the torch until it is plastic enough to bend easily. After shaping in this manner, the pieces are welded together to form the completed article.

SUMMARY

1. Every technique you learned in oxyacetylene and arc welding can be adapted to art metal work.

2. Small oxyacetylene torches are used for small and delicate welds and brazes.

3. Stick figures and decorative objects are easy to make and best to begin with. Stick figures are made by bending and welding or brazing-filler rods into cartoon outlines.

4. Almost any type of weldable or brazeable metal can be used for art metal work, provided the proper filler rod is used. This includes tin cans and junk material that can be fashioned into small art objects such as flowers, birds, animals, and butterflies.

5. In straight welding we are concerned with regular beads. In art metal this is not always true. We frequently drip the melted rod to create lumps and distorted forms for texture and form.

6. Wrought iron is a regular material for decorative work. Its strength and malleability permits its use in many forms.

7. Mild steel is used frequently for commercial ornamental iron work and can be adapted for many designs. Steel, however, is not as resistant to rust as wrought iron.

Figure 19-16. A wrought iron fence adds security and neatness to a yard. This wrought iron fence was welded together in sections after being constructed in shorter segments.

HOME WELDING

A welding setup is very handy in the home garage. Many students start planning one as soon as they gain some measure of proficiency. You will find that many professional welders also have a home workshop. Unless you own the shop or factory where you work, you will not be able to bring your own work in to weld on the company's time—like a man's wealth, the boss's equipment is something you cannot take with you.

By the same token, a home welding setup cannot have—nor does it need to have—all the expensive equipment you will find in a professional shop (Figure 20-1).

FIRST, IS IT LEGAL?

In rural areas you are pretty much on your own, but in cities you have to consider zoning laws and city ordinances before you set up a shop in your garage. This is especially true if you expect to do any work for hire. Also, you had better take a look at your fire insurance to see if any such action will void your policy.

The restrictions may depend upon the kind of equipment you intend to use. There are small welders and oxyacetylene kits that are Underwriters Laboratories approved and which can be used for small jobs around the house or in a garage. These should not come under the heading of professional or commercial equipment, yet they can do a professional job in the hands of a trained operator. With them you can repair tools, fix

Figure 20-1. Here is a good example of a home welding shop. There is an iron top table, storage space for the welding machine, an electrode holder, vise, extra tools, and a convenient electrical outlet.

plumbing, straighten fenders, patch rain gutters, and perform many other money-saving jobs (Figure 20-2).

SAFETY FIRST

The usual safety precautions must be taken. Ventilation must be adequate, fire hazards removed, curtains put up, fire extinguishers and fire blankets available, and the electrical current must be sufficient.

Figure 20-2. Many practical or money-saving jobs can be done in a home welding shop, such as this homemade storage rack with a ladder for easy access. A home welding setup permits you to try many such creative ideas.

Figure 20-3. A shop table can be made with a 1/8 to 1/4 in. thick steel plate top and pipe legs. A vise is convenient for holding material for welding, such as this basketball hoop.

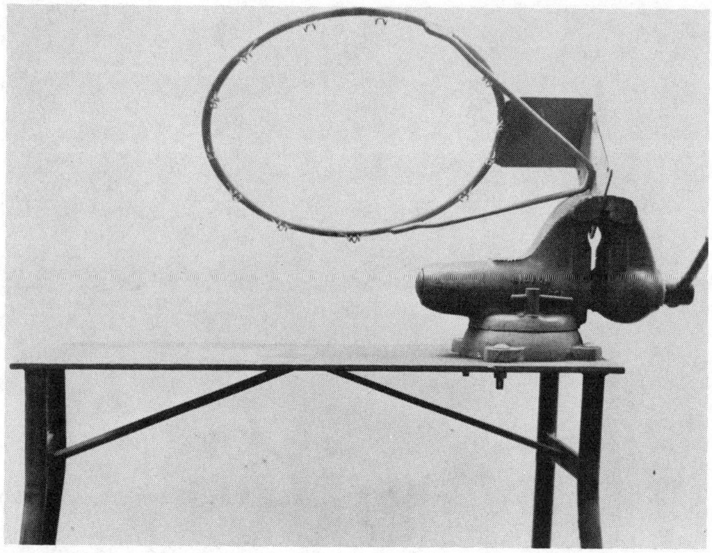

Figure 20-4. The portable electric grinder is very handy for cleaning up weld joints or grinding and shaping metal. It uses either a disk or a grinding stone.

Figure 20-5. This grinder is mounted on a pedestal and stand that is anchored to the floor. A water holder (b) is used to quench hot metal. The tool rests (a) are safely positioned.

For the shop itself you should have a table with an iron plate top. You can make one yourself, using 1/8- to 1/4-in. plate with pipe legs and braces. Such a work table will not catch fire from torch, arc, or hot materials. It will also be strong enough to withstand considerable hammering. A large vise is also a handy tool for securing metal during welding or straightening.

OTHER EQUIPMENT

In addition to the actual welding equipment, which will be discussed later, there are a number of items that will prove handy in your home welding shop. One of these is a combination grinder and buffer. You can use it for beveling metal, smoothing weld beads where this is desirable, removing nicks and burns, and for polishing.

It is easy to make a grinder of this type. Many old washing machine motors can be converted to run grinder or buffing wheels. The motor can be secured to a table, or pedestal-stand mounted (Figure 20-5). The pulley is removed from the shaft and replaced with a small sleeve attachment that can be obtained at a hardware store. This attachment fits over the shaft and has a threaded portion that holds the grinding or buffing wheel.

Clamps or metal gripping vise grips to lock or position pieces of metal in place are also very helpful. These devices are like a third hand.

STORAGE

Storage space is always a problem in home welding setups. A lot of your work will be with scrap metal, and it is frequently a problem where to keep it. The storage place should be out of the weather to prevent rusting, and strong enough to support the weight of the metal.

Wall space is an ideal storage place. Separate the wall into 1- or 2-ft areas and use pegs or 2 x 4 brackets to form the sides of each space. Then the larger and longer pieces of round, flat, or angle stock can be stacked between the pegs.

Wooden boxes about the size of apple crates are good for saving small pieces of steel, arranged in common thicknesses and shapes.

Figure 20-6. This photograph shows the arrangement for the motor shaft adapter used for installing a grinding wheel.

Figure 20-7. This is a homemade buffer wheel. The motor has a pulley that uses a belt to drive a shaft. The shaft is mounted in a bracket and has two adapters on either end. Each end of the shaft has a wire buffing wheel mounted on it.

Figure 20-8. Metal storage is always a problem, but having an organized method helps make the job a lot neater. These handy racks store round stock.

Figure 20-9. These storage cabinets are very handy for making your extra pieces or leftovers easy to locate. The tool box is ideal for keeping your tools safe and orderly.

Electrode Storage

Shelf space must be provided for your electrodes and filler rods. An oxyacetylene rod is usually 30 in. long; arc welding electrodes are usually 14.

Pieces of 3- to 5-in. diameter pipe welded to a piece of channel iron for a base make an excellent storage place for arc electrodes. These flux-coated rods are susceptible to moisture. They come in sealed containers and should be left in containers until needed to avoid absorbing moisture from the air. Low-hydrogen electrodes must be stored in an oven at about 280°F to control their special low moisture content.

Oxyacetylene filler rods can be organized into types, that is, brazing or welding, and further distributed by diameters.

Of course, if you buy only what you need for each job, the problem of storage is simpler. Some welding supply houses will sell individual rods and electrodes. Others will package them in one-dozen lots, but some refuse to break open sealed cans for small lot sales. You save more money, naturally, by buying in larger quantities.

AIR FLOW

We have discussed ventilation from the standpoint of safety and the dispersal of toxic fumes. If you are working with oxyacetylene, there is an additional ventilation problem. A direct flow of air from a window box fan, an open garage door, or other sources may be great for keeping the air flow moving around you, but a direct air current can blow your oxyacetylene flame, making your flame control more difficult. The air flow can also cool your metal, which makes it harder to control the molten puddle.

This problem is solved by keeping the air flow above or to the side of your work.

HOME WELDING EQUIPMENT

Equipment for home use may be bought or rented. Whether you buy or rent and the type you need are determined by the use you have for the welding outfit.

If you decide to buy, there are a number of welders available for home use in a price range to fit any pocketbook. Arc welders range from $19.95 to several hundred dollars. You will find them for sale in hardware stores, auto supply houses, and at welding supply dealers. The cheaper models do not have good arc control. The higher priced machines offer increased stability in the quality of the electrical frequency. To the beginner, this means that the arc can be struck easier and the arc controlled better on the more expensive machines.

Duty Cycle

Another factor to look for in choosing your home welding machine is the "duty cycle." This tells you the amount of work the arc welder will perform at full rated output. For example, if the duty cycle is rated at 30 percent, it means that the arc welder can put out its full rated output for 3-min out of a 10-min span. If you try to weld at full output amperage for more than the rated time (3 min), the machine will overload and shut down.

Duty Cycle Variations. Most of the higher priced machines are rated 40 to 60 percent duty cycle. The lower priced models are rated about 20 percent. Most of your welding will take place with 150 amps or less. This means that your rated machine output should be 200 to 250 amps. This provides you with maximum coverage for welding jobs, plus reserve capacity for duty cycle protection. You will get this capacity in the higher priced machines.

If you drop below this level machine, you must sacrifice something. For example, you may find a machine that uses 100-V input for a rated output of 180 amps and a duty cycle of 30 percent. This is adequate, but you will notice a difference in the output of machines with 100-V input as compared to those using 220-V input.

Voltage input and duty cycle will also apply if you decide you want a machine with DC polarity features. No matter what you buy, you must always balance need against cost and desire against good judgment. The ability to change polarity provides greater flexibility in penetration, heat control, and electrode selection but will add to the cost.

Buying Used Equipment

The simple AC "buzz" box or transformer type of welding machine has few moving parts and is easily serviced. Normal maintenance involves cleaning and replacing the contact switches. It is not uncommon for this type welder to give 10 to 15 years service.

The combination of AC-DC welding machine with the rectified transformer current is also durable. The motor-driven DC generator type has more moving parts and may require additional service for the generator portion of the unit.

Cables, electrode holders, and ground clamps are the most frequently damaged parts. They are easily replaced.

CHECK YOUR HOUSE WIRING

Before installing any arc welding equipment in your garage, check your electrical system to make sure it will carry the load. Electrical circuit breakers and fuses must be sufficient to protect the line from overloading.

You must also be careful about the use of extension cords. Depending upon the quality of the machine, operation may be affected by the length of your extension cord or cords. Always use a heavy duty grounded extension cord and keep the length to a maximum of 25 ft.

If you need extended lengths of cord, it may be wise to consider adding an extra electrical outlet closer to your work. You will, of course, have to follow the electrical code in effect in your area when making these extra connections.

OXYACETYLENE EQUIPMENT FOR HOME USE

Where the higher heat of arc welding machines is not desirable, you might want to consider a home oxyacetylene setup. The arc machine is more convenient, compact, and you do not have to worry about cylinders of oxygen and acetylene. Oxyacetylene can be used for art metal work or for easy portability without depending upon extension cords. It will also operate in areas far removed from electrical outlets, which is certainly a major advantage. Oxyacetylene also creates less splatter.

Oxyacetylene outfits are not overly expensive. A torch, an assortment of tips, and two regulators may be obtained for about $175 (1975 prices) for the standard size. This does not include the oxygen and acetylene cylinders and gases or a hand cart.

Welders doing an appreciable amount of work

Figure 20-10. Oxyacetylene is suitable for home garage use. This welder is using a small tank oxyacetylene system in his garage. He is finishing the frame for an automobile engine stand.

generally do not buy cylinders. They are obtained full of gas from suppliers and returned when empty. There is usually a demurrage charge if you do not return the empties within a specified time, generally 30 days. For this reason those using a small amount buy their cylinders. Major suppliers have 80-cu ft oxygen and 60-cu ft acetylene cylinders for sale. These small cylinders are very popular.

Big or Small?

The basic outfit mentioned above is for general-purpose work. If you are doing a lot of work, you might want a heavy-duty outfit. It will include the torch, tips, and hoses of the regular outfit, but the regulators will be double stage instead of single. The double-stage regulator gives a more steady gas flow.

There is also a minitorch set at about half (or less) the cost of the regular set. In some sets the torch is not much larger than an oversize pencil. These minitorches are ideal for working on small art metal jobs or for electronic work.

The choice of an outfit must be based upon its intended use. This will include the type of metal—light or heavy—how much work you intend to do, and whether you will do all your welding in the garage or need portability.

Other Considerations

Beyond the basic consideration of use, there are some other things to think about in buying oxyacetylene equipment. If you buy little-known brands, you might have trouble getting accessories or replacement parts. Ensure that such material is available before making a purchase. Another point to consider is the "feel" of the torch. How does the weight and balance feel? Can you hold it comfortably? Also check to see that the manufacturer uses a standard numbering system for his tips.

Buying Used Oxyacetylene Equipment

As mentioned earlier, acetylene and oxygen cylinders are not usually sold but remain the property of the supplier. This system facilitates checking containers and ensuring that they are safe to use. However, you will sometimes find them for sale on the second-hand market. It is important they be of approved construction, otherwise you may find suppliers refusing to refill them for you. Used regulators must be tested—they must work smoothly, control the gas flow accurately, and the valves must seal tightly.

The hoses should be inspected for leaks. Torch tips and accessories should have good tip surfaces. Worn O rings are no particular problem since they can be easily replaced. The same is true of the cart. A few minutes with the torch and it should be good for another 10 years.

Tank Sizes

Tanks are sized together in certain basic sizes or sets. The size you choose will be determined by the amount of use you expect to have for the unit. The smaller tank set requires less space to store and is easily moved around.

SOLID OXYGEN WELDING OUTFIT

A small welding torch outfit recently appeared on the market that greatly simplifies gas fusion welding. It does not require all the components of an oxyacetylene setup, nor does it have a regulator. It is marketed under the trade name of *Solidox.* It uses propane gas in the place of acetylene, and the oxygen is supplied by burning specially prepared pellets (Figure 20-11).

The unit blends the two gases that are dispensed through hoses to a miniature torch. There is no regulator. The control valve on the propane tank is adjusted to control the flame characteristics.

The torch is lighted in a unique manner. The color-coded end of the oxygen pellet is lighted and placed in a sealed container, the propane valve is opened (mixing is automatic), and the mixed gases are ignited with a striker.

Each oxygen pellet will burn for about 8 min, and then a new pellet has to be installed and relighted. The propane is supplied in small camping size cylinders and

Figure 20-11. Portable mini-welding kits are handy for making home or farm repairs. Here a kit is used for brazing a broken lawnmower casting.

can be shut off with a valve. The oxygen pellet will continue to burn but can be extinguished by removing it from the holder and cutting off the burning section, saving the stub for relighting later.

Regular welding and brazing rods may be used. The oxygen pellets, good for 65 min total burning, cost—at present prices—about \$4 to \$4.50, which makes the system rather expensive to operate. However, the lightweight portability of the unit and its ease of operation make it ideal for the home welder who has

Figure 20-12. One type of portable gas welding kit uses propane gas in connection with solid oxygen pellets for fuel. The kit weighs 7 lb and is capable of handling small welding, brazing, and soldering jobs.

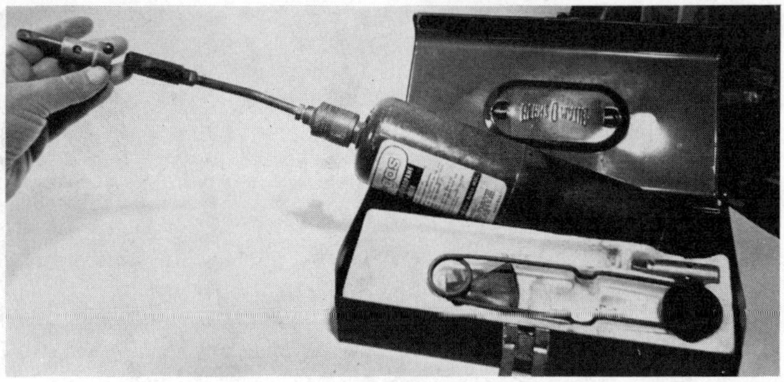

Figure 20-13. A propane torch kit is a useful heating tool. It can be used for many repair jobs including those involving solder, for removing paint, or for bending.

only an occasional job to do. The torch is unique in its compactness. It has an adequate heat output that can be used for both light- and heavy-duty welding, brazing, or cutting.

BERNZOMATIC TORCH

These small kits include a propane cylinder and various components for soldering, heating, or flame spreading. They are handy for heating light gauge metal, soldering rain gutters, or other sheet metal applications.

Included are an open tip, a closed tip, and a flame spreader attachment. Either the open or closed tip can be used for soldering, depending upon the type of job. The closed tip is used for electrical wires, and the open flame works a little faster on sheet metal.

The flame spreading attachment is used to make the flame wider. It can then be used to soften paint so it can be scraped off with a putty knife. A slightly discolored surface will not cause any damage. The surface can be readily repainted with a minimum of sanding.

Figure 20-14. This abrasive cut-off wheel is a carborundum disk that is belt driven by a motor mounted underneath. This machine cuts metal much faster than a hacksaw.

ABRASIVE CUT-OFF WHEEL

This quick-cutting tool is used to take the place of a hacksaw. The carborundum blade quickly slices through metal. The cutting disk is belt-driven by a pulley on a motor. As the disk is brought down against the stock the disk cuts through the metal. It is more noisy than a hacksaw, but does the job much quicker. It is not a common household item, but it is being more widely used because of its speed and accuracy in the cut.

SUMMARY

1. Welding does not need to be confined to a professional equipment to oxyacetylene minitorches around the house, garage, or farm for a good welder.

2. Home welding setups may extend from full-fledged professional equipment to oxyacetylene minitorches and propane-fueled kits.

3. The type of equipment for home welding depends upon the type and quantity of work you expect to do. A small, lightweight outfit like the Solidox, which uses propane and solid pellet oxygen, is adequate for an occasional job.

4. A garage or home welding setup must be in conformance with local and underwriters regulations and codes. Fire hazards must be carefully checked because this is *your* house.

5. The work area must have ample ventilation to carry away toxic fumes, but a direct breeze may fan the oxyacetylene flame and make it difficult to weld.

6. Storage space must be provided for metal, rods, electrodes, and welding equipment.

7. Both arc and oxyacetylene equipment are suitable for home welding. Arc is the most convenient because it requires only a machine, cables, electrode holder, and ground clamp.

8. Low-priced arc welders have poor provision for stabilizing the electrical frequency.

9. Duty cycle is a factor to consider in choosing an arc welder. This is the amount of work the welder will perform at full-rated output.

10. Small oxyacetylene outfits are available for home welders who do not need full professional size equipment. Oxyacetylene has the advantage that it can be used where there are no electrical outlets, which are necessary for arc welding.

11. Specialized light-duty equipment is available, such as the Solidox, which uses propane gas and solid pellet oxygen.

12. Propane torches alone, used without oxygen, are good for soldering.

NEW WORDS AND TERMS

Abrasive Cut-Off Wheel A carborundum disk that is motor driven; it cuts metal faster and neater than a hacksaw.

Bernzomatic Torch A propane torch with accessories to heat, solder, or remove paint.

Duty Cycle A term to indicate the ability of an arc welding machine to work under a load. It is defined as the length of time a machine can produce its fully rated output during a 10 min span.

Solidox A trade name for a solid oxygen process weld-braze outfit. It burns propane gas with oxygen from solid pellets.

WHERE DO YOU GO FROM HERE?

This book is an introduction to welding. Its purpose is to cover only the basic principles of oxyacetylene and shielded metal arc welding. Once you have mastered these basic techniques you can and should move ahead in your profession. Welding, although an ancient profession, is new in the sense of its modern development. It is also growing and changing as welders confront new problems of difficult-to-weld alloys and the challenge of mass production.

TIG WELDING

One of the more advanced welding processes you will want to learn is Gas Tungsten Arc Welding—or TIG (Tungsten Inert Gas) welding as it is almost universally called. As you learned in connection with arc welding, oxygen and nitrogen in the air react unfavorably upon molten weld puddles and the cooling bead. This oxidizing reaction produces weak, porous welds. In regular arc welding the flux burns to produce a gas that partially shields the puddle from the atmosphere, then the flux produces a slag which covers the bead to protect it while the bead cools.

This flux shield is not sufficient protection when we weld complicated alloys where maximum strength is essential, such as in aviation and aerospace equipment.

In TIG welding we do not use a consumable electrode as in shielded metal arc welding. The TIG electrode is tungsten and is used only for creating heat. A blast of inert gas from a tank to a flow cup blows around the electrode to create a shield against the atmosphere. Argon or helium is used for the inert shielding gas. It is called inert because it does not react with the hot metal as does oxygen and nitrogen.

TIG Welding Machines

TIG welding differs from regular arc welding only in the use of a nonconsumable electrode and the use of shielding gas to protect the metal. The arc welding machines may be the same or may be specially built for the TIG system. They may be AC or DC, and you may use different polarities with the DC current. A cylinder and hose arrangement feeds the inert gas to the electrode holder. A few other machine switches or controls are also different.

MIG WELDING

Another advanced technique is MIG welding. This is also a form of shielded gas welding. Although universally known as MIG to welders, the correct name is Gas Metal Arc Welding. Manufacturers have their own trade names for it. For example, the Hobart company calls its MIG system Micro-wire Welding. Linde has one named Sigma Welding, and so on.

MIG welding uses a shielding gas, as does TIG, but the major difference between the two is in the electrode and the way filler metal is deposited. MIG uses a consumable wire as the electrode. This wire is reeled through the torch tip, serving both as an electrode and as filler metal.

Advantages of MIG Welding

MIG welding has deep penetration and a high speed of application—both very desirable in production welding. Since the filler is fed from a reel, the operator does not have to manipulate the filler rod. No flux is needed. This eliminates the formation of slag, which speeds up cleaning the weld bead.

A MIG unit consists of a welding machine to supply the proper current as in all arc welding processes. In addition, there is a reel for the wire electrode, a motor to feed the wire through the system, a control unit, a gas cylinder to provide the shielding gas, cables, and the welding gun. The gun takes the place of the clamp in regular arc welding. It is held by the operator and feeds the wire and gas to the weld puddle.

Figure 21-1. The MIG welding gun features a consumable electrode. Inert gas blows around the electrode to provide a shield for the puddle.

PRODUCTION WELDING

The type of welding you have learned to this point requires an individual welder working by himself to create a single job at a time. There is plenty of work for such operators, but in mass-production factories individual welding setups are too slow and costly. Big automobile and aerospace plants (Figure 21-2) have developed automatic production line welding to turn out welded products in quantity.

Many production welding machines do not require trained welders but use welding machine operators. This is not to imply that such operators lack skills or do not require proper training. It means that many of the technical operations that an individual operator must do for himself are done automatically by the machine. A good machine operator would not necessarily be able to leave his production welding line and pick up a manual oxyacetylene torch and run a good bead without extra training.

Figure 21-2. These General Motors Vega bodies pass through an automatic welding assembly production line. Ninety-five percent of the 3,900 welds on each car body is made by automatic machinery. (Photograph courtesy of General Motors.)

Production welding machines are excellent for their purposes: mass production. They cannot be used in the field, on top of skyscrapers, for small jobs, or for anything except what they were designed for. Thus they can never replace the individual welder with his torch or arc.

ELECTRON BEAM WELDING

Some of the new alloys developed for the aerospace industry are difficult to weld by ordinary processes. One solution to the problem of how to handle this type of metal is electron beam welding, which uses a stream of electrons instead of an electric arc. An electron gun emits a stream of electrons from a heated filament. Electrons can be focused the same as light rays are focused, by a glass lens, except that the lens is a magnetic field. The electromagnetic lens in an electron beam welder bends the electrons to a focus. This produces an intense hot spot, just as the sun's rays are focused to a hot burning spot by a magnifying glass.

The intensely hot beam makes a weld with a minimum of distortion and in extremely small sizes. The electron beam welder is used mainly in the aerospace industry for high-quality joining of exotic metals. Its major disadvantage is that air stops the electron flow and the entire process must be carried on in a vacuum chamber. This limits the size of the material to be welded.

LASER WELDING

Another unusual welding process is laser welding. Laser light is a highly concentrated light that shines in a single direction instead of all directions as ordinary light does.

A narrow beam of laser light can be made to produce heat up to 10,000°F. The resulting beam can be made so small that it has been compared to a white-hot needle. Thus the laser can weld metal without disturbing the adjoining metal. It can also join dissimilar metals, such as copper to aluminum, and stainless steel to tungsten. However, laser welding is limited because of its shallow penetration. In spite of this limitation, it is finding application in the electronics industry for joining small, difficult-to-weld alloys.

YOUR FUTURE IN WELDING

Despite the development of automatic processes, there is now and always will be a place and job for the individual welder. According to one survey report, the nation will require 30,000 new welders in the next 4 years. Basic instruction in oxyacetylene and arc welding is sufficient to qualify you for many jobs, but if you wish to advance in your profession, you must study advanced welding techniques. You will need a deeper knowledge of metallurgy and must learn how to read blueprints and welding symbols so that you can follow precise instructions. You must also learn individual techniques for handling difficult-to-weld metals.

Actually, your future in welding is up to you. You may go to work in a shop, a factory, or on a project in another country. There are jobs for you in bridge building, skyscraper construction, assembling huge machines (like tanks and earth-moving equipment), and so on through a seemingly endless list. No man, they say, is indispensable, but the competent welder in our modern society comes as close to it as anyone.

You have picked a sometimes glamorous, sometimes adventurous, and always interesting profession. Your future in it is limited only by your imagination and your desire to work hard and get ahead.

Good luck!

INDEX

AC (alternating current), 228
AC welders, 228–229, 251, 262
Acetylene, 7, 16, 50–51, 68
Acetylene cylinders (tanks), 52–53, 66
Acetylene generator, 51, 68
Alloy metals, 36
Aluminum, soldering, 191
American Welding Society, 103, 110
Amperage settings (arc welding), 243–244, 249
Annealing, 33, 34
Arc, formation of, 224
Arc blow, 231–232, 257
Arc crater, 15, 16, 227, 268–269
Archimedes, 44
Arc length, 231
Arc welding, 7, 12, 15, 224–237, 255
  striking an arc in, 14–15, 251, 254–256, 263
  equipment, 225
    types, 228, 229
Art metal welding, 278–291
  equipment, 280
  materials, 283
  types of work, 281
Automatic cutting torch (*see* Flame cutting machine)

Backfire, 54, 98, 123
  causes of, 98
Backhand welding, 154–156
Bead:
  factors in quality, 111, 119, 267
  forming, 111–119, 227–236
Beveling, 75–76, 83, 102
  with flame cutting torch, 206–207
Blowholes, 269, 270
Braze welding, 145, 164, 167–178
  definition, 106, 110
  filler rods, 105, 168
  techniques, 175–177
Brazing, 36, 164, 165–178
  advantages, 164
  capillary attraction in, 105, 173
  flux use in, 105, 168–171
  techniques, 165–167, 174–175
  rods used in, 105
Butt joints, 10 74–75, 261
Butt weld gap, 121

Cadmium fumes poisoning, 82
Carbide, use in making acetylene, 51
Carburizing flame, 95, 181
Casehardening, 147–148
Cast iron, 37
Chip test, 42
Circuit, electrical, 225–227
Clamps, to prevent distortion, 82, 120, 145
Cleaning metals, 70–77
Color code, electrodes, 242
Compressive strength, 141
Copper pipe, soldering, 190
Corner joint, 79, 124
Crater (*see* Arc crater)
Critical points, 31, 45
Critical range, 31, 140
Cylinders (acetylene and oxygen), 52–53, 59–60, 66
  care of, 55, 66
  inspection, 55
  lubrication of valves, 55

DC (direct current), 228
DC welding equipment, 229, 253, 262
Destructive testing, 131, 265, 273
Dipping (filler rod), 117, 151, 159
Double-vee butt joint, 75–76
Ductility, 142–143
Duty cycle, 298–299

Edge joint, 77
Elasticity, 141
Electrodes, 12, 22, 239, 253
  coding of, 239–240, 241

color coding, 242
diameters, 243, 249
storage, 249, 297
types, 239, 249
Electrical circuit, 225–227
Electrical resistance, 224, 236
Electron beam welding, 9, 310
Expansion of metal, 146
Extra high carbon steel, 39

Face shields, 27
Fatigue strength, 141
Ferrous metals, 37–38, 45
Filler rods, 12, 102–106, 108–110
Fillet welds, 72, 125–126, 130
Fillet weld testing, 275
Filter, lens, 24–26
Fishmouth (pipe contour cutting), 135
Flame arrester, 54
Flame cutting, 194–196, 201–205
Flame cutting machine, 217–219, 222
Flame:
adjusting, 94–97
focusing the cone of, 96
oxyacetylene, 93–97
types, 95-96
Flange weld, 77-78
Flashback, 123
Flow point, 173
Flux, 15, 105, 245–246, 249
iron powder flux, 247–248
role in arc welding, 245
use in brazing, 168–171
use in soldering, 182, 183, 186
Forge welding, 5
Fracture test, 42

Gap, 8, 15, 16
Gloves, 22
Glycerine, as lubricant, 66, 69
Goggles, 21–23, 28
Grain, effect on hardness, 30, 31, 32, 45–46, 138
Grain in metal, 30, 31, 45–46, 138, 148
Gray cast iron, 37
Guided bend test, 273

Hard facing, 105, 110
Helmet, welding, 24–26
High carbon steel, 35–36, 38
Holes, how to cut, 212, 216, 221
Home welding, 292, 305
Home welding shop, 293, 297–298
ventilation, 297–298
Horizontal welding, 157–158
Hoses, welding, 67, 84

Identification of metals, 39–45
Impact strength, 142
Inclusions, 269
Inspection of bead, 265–276
Interstate Commerce Commission, 54–55

Jigs, 81–82, 145
Job opportunities, 311
Joints in welding, 71–80

Kerf, 203
Keyhole technique, 75

Lap joint, 72, 128
Laser welding, 310
Leaks, test cyclinder for, 55, 66, 69, 92
Low carbon steel, 30, 31, 32, 38

Macroinspection, 272
Malleable iron, 37
Manifold, 54
Medium carbon steel, 38
Melting points of metals, 32, 33
Metallurgy, 30–45, 138
Metals:
cleaning, 70–77
expansion of, 146
ferrous, 37–38, 45
grain, 30, 31, 45–46, 138, 148
identification of, 39–45
melting points, 32, 33
nonferrous, 37
MIG welding, 308
Multiple pass welding, 128, 159

Neutral flame, 95
Nondestructive testing, 265, 272
Nonferrous metals, 37
test for, 43

O rings, 89
Out-of-position welding, 150–162
Outside corner weld, 79–81
Overhead welding, 150, 159
Oxidizing flame, 95
Oxyacetylene equipment for home use, 300, 302, 304
Oxyacetylene flame, 93–97
adjusting, 94–97
focusing the cone of, 96
types, 95–96

Oxyacetylene test for metals, 43
Oxyacetylene welding, 7, 10, 50–69, 84-101
Oxygen, 50, 56–57, 68
Oxygen cylinders, 56, 58, 59, 60
Oxygen hoses, 67

Patterns, welding, 257–259
Peening (*see* Stress)
Penetration, 257–268
Pig iron, 37
Pipe cutting, 213, 221
Pipe-to-pipe welding, 133–134
Pipe-to-plate welding, 134
Pipe welding, 132
Plate metal, 111
Plug welds, 77
Polarity, 227, 229
Polarity switch, 230
Powder cutting, 220–221, 222
Production welding, 309
Protective clothing, 14, 21, 28
Puddle, carrying the, 12, 115–116
Puddle heat, judging, 120

Quenching, 34–35, 46, 138

Radiation, 21, 28
Regulators, 64–91
  adjustment, 91, 112–113
  attaching, 91
  single-stage, 64
  two-stage, 64
Resistance, electrical, 224, 236
Restarting weld bead, 261, 264
Reverse polarity, 230, 236
Ring test, 42
Ripple bead, 15
Root pass, 73
Rosebud tip, 89

Safety, 18–28, 66–159, 232, 234
Scarfing tip, 90
Sheet metal, 111
Shop, home welding, 293, 297
Silver soldering, 105, 181
Slag, 15, 205, 246–247
Soldering, 105, 165, 181–193
Spark tests, 40–42
Specific gravity tests, 44–45
Splatter, 274
Sticking rods, 120
Straight polarity, 230–236

Strain, 141
Strap welds, 81
Stress, 140–141
  removing, 144
Striker, 93–94
Stud welder, 8
Sweating (in soldering), 187–188

Tack welds, 81, 123
Tensile strength, 141
Testing welds, 130, 161
Thermit welding, 6
TIG welding, 307–308
Tinning, 176
Tips:
  cleaning, 97, 200
  fitting, 96
  flame cutting, 89, 199
  sizes, 86–88
  torch, 86, 89, 97, 200
T joints, 73, 262
Torch, welding, 84–101, 114
  flame cutting torch, 89-90, 194, 197–199
  preparing torch for work, 92, 94
  shutting down torch, 97–98
  summary of operational procedures, 100
    (*See also* Flame)
Torsional strength, 141
Two-stage regulator (*see* Regulators)

Undercutting, 127, 257, 270
Used equipment, buying, 299, 301

Vee joints, 145
Ventilation in shop, 297–298
Vertical down weld, 156
Vertical welding, 150–153, 156
Visual examination, 265, 271

Warpage, control of, 122, 140, 145, 148
Weaving (*see* Patterns)
Welding:
  description, 2, 9
  history, 4–8
  job opportunities in, 311
Welding hoses, 67, 84
Welding patterns, 257–259
Welding torch, 84–101, 114
Whipping, torch, 117, 260
White cast iron, 37
Wrought iron, 289, 290

Zinc fumes, danger of, 172